T0257806

Climate Change:
Researches in Regional Responses

Climate Change: Researches in Regional Responses

Edited by **Daisy Mathews**

New York

Published by Callisto Reference,
106 Park Avenue, Suite 200,
New York, NY 10016, USA
www.callistoreference.com

Climate Change: Researches in Regional Responses
Edited by Daisy Mathews

International Standard Book Number: 978-1-63239-115-5 (Hardback)

Contents

Preface

This book has been a concerted effort by a group of academicians, researchers and scientists, who have contributed their research works for the realization of the book. This book has materialized in the wake of emerging advancements and innovations in this field. Therefore, the need of the hour was to compile all the required researches and disseminate the knowledge to a broad spectrum of people comprising of students, researchers and specialists of the field.

This book presents a descriptive account on the concepts of climate change and discusses research-focused information regarding its regional responses. Researchers from across the globe have contributed valuable information in this insightful book. This book discusses various topics related to this field like forecasting, measurements as well as observations of meteorogical visibility, grids in numerical weather and climate models, etc. Updated in-depth information has been compiled in this book. The primary aim of this book is to serve as a valuable source of information for researchers, scientists, students, etc. interested in studying about the concepts of climate change and related topics.

At the end of the preface, I would like to thank the authors for their brilliant chapters and the publisher for guiding us all-through the making of the book till its final stage. Also, I would like to thank my family for providing the support and encouragement throughout my academic career and research projects.

Editor

Weather Forecasting

Ensemble Forecasting

Alfons Callado, Pau Escribà,
José Antonio García-Moya, Jesús Montero,
Carlos Santos, Daniel Santos-Muñoz and
Juan Simarro

Additional information is available at the end of the chapter

1. Introduction

The atmospheric movements can be described by non-linear differential equations that unfortunately have no analytical solution. The numerical methods to solve the atmospheric non-linear differential equations have been developed in different stages. During the 50s, Charney, Fjørtoft and von Neumann (1950) made a 24-hour forecast of 500 hPa geopotential height using a bidimensional quasi-geostrophic equation. After that, in 1956, Philips showed the close relation between cyclone dynamics and the global circulation using a 2-layer model.

At the beginning of the 70s, the global circulation models emerged (Lynch, 2006). These models are based on a set of non-linear differential equations, which are used to approximate the global atmospheric flow, called primitive equations. During this stage the full primitive equations were implemented without any quasi-geostrophic approximation (Williamson, 2007).

During the 80s, the regional and mesoscale numerical models appeared (Athens &Warner, 1978; Mesinger et al., 1988). The evolution of the models is a direct consequence of the increase of computer resources, and the improvement in observational networks and assimilation methods. This evolution has extended the knowledge on the dynamics and atmospheric microphysical processes.

The last period of the numerical weather prediction was initiated in the 90s. The atmosphere-ocean and atmosphere-ocean-soil coupled models, and the spatio-temporal high resolution models allowed the development of analysis and diagnostic techniques for the weather forecasting (Mechoso & Arakawa, 2003).

Until then, the numerical prediction models' philosophy was based on the deterministic atmospheric behavior. That means, given an atmospheric initial state its evolution can be numerically predicted to give a unique final state. Consequently the efforts of the scientific community were focused on producing the most accurate prediction (Tracton & Kalnay, 1993).

Nevertheless, the formulation of models requires approximations due to unknown variables or known process that cannot be explicitly resolved using the spatio-temporal resolutions a model works with. These processes must be parameterized and this fact generates errors associated with the parameterization used in the model. Although the model could perfectly simulate all the atmospheric processes, it would be impossible to determine a realistic initial state description for all resolutions and in all places using the available observational data (Daley, 1991). Lorenz (1963) showed that small variations on the model initial conditions do not produce a single final solution, but a set of different possible solutions. That is why the predictability of the future atmospheric states is limited in time: the initial condition errors are amplified as the forecast period grows (Lorenz, 1963, 1969).

The traditional deterministic approach gave way to a new paradigm, with richer information than a single solution of the future atmospheric state. The new paradigm includes quantitative information about the uncertainty of the predictive process. The atmospheric non-linear behavior, consequently chaotic, must be treated now in a probabilistic way (Lorenz, 1963).

The improvement of numerical models will permit a better characterization of the atmospheric processes but the models will always have some limitations related to the scales of the simulated processes and the approximations made to solve numerically the equations. Another limitation of the numerical forecasting methods is the lack of observational data with high enough resolution to properly describe the initial state.

Nowadays the observational methods, the assimilation strategies and the own characteristics of the numerical models have inherent limitations that generate uncertainty in the estimation of the possible future atmospheric states. The uncertainty is amplified when the forecast period grows and when the resolution increases. Thus, the probabilistic approach seems an ideal strategy to characterize forecast uncertainty.

The atmospheric state cannot be exactly known. The analysis data always contain an error that only can be estimated. The inaccurate determination of the real atmospheric state drives to the existence of a great number of initial conditions compatible with it. A single model only provides a single solution of the future atmospheric state. The generation of multiple forecasts starting from slightly different but equally-probable initial conditions can characterize the uncertainty of the prediction (Leith, 1974).

The generation of equally probable forecasts starting from multiple realistic initial conditions introduces the probabilistic forecasting concept. A practical approximation to probabilistic forecasting based on meteorological models is the so called ensemble forecasting. The Ensemble Prediction Systems (EPS) are used operationally in several weather and climate prediction centres worldwide. The European Centre for Medium-Range Weather Forecasts (ECMWF; Molteni et al., 1996) or the Meteorological Service of Canada (Pellerin et al., 2003), among others, produce routinely ensemble predictions. These predictions have been demonstrated to be extremely useful on decision making process.

The EPS is a tool for estimating the time evolution of the probability density function viewed as an ensemble of individual selected atmospheric states. Each of these different states is physically plausible. The spread of the states is representative of the prediction error (Toth & Kalnay, 1997).

Several techniques for constructing the ensemble have been developed and applied. One of the first methods proposed for generating an ensemble of initial states is the random Monte Carlo statistical methodology. It was proposed by Leith (1974), Hollingsworth (1980) and Mullen and Baumhefner (1989), among others.

Perturbative methods that depend on the atmospheric flow are also used. These strategies are based on the generation of perturbations in the subspaces where the initial condition errors grow faster. The breeding vectors (Toth & Kalnay, 1993, 1997) or the singular vectors (Buizza & Palmer, 1995; Buizza, 1997; Hamill et al., 2000) are remarkable examples.

There are other perturbative methods that consider the model sub-grid scale errors by means of varying model physical parameterizations (Stensrud et al., 1998; Houtekamer & Mitchell, 1998; Andersson et al., 1998) or using stochastic physics (Buizza et al., 1999).

The combination of multiple model integrations initialized by multiple initial conditions determined by different analysis cycles is another strategy to generate ensembles. Using different assimilation techniques allows characterizing the uncertainties associated to the initial condition and the uncertainty associated to each model (Hou et al., 2001; Palmer et al., 2004). Finally, taking different global models as different initial conditions has been found to provide better performance than any single model system (Kalnay & Ham, 1989; Wobus & Kalnay, 1995; Krishnamurti et al., 1999; Evans et al., 2000).

The technique based on the use of multiple limited area models (LAM) and multiple initial conditions coming from several global models combined with advanced statistical post-processing techniques (Gneiting & Raftery, 2005a) has been tested in the National Centres for Environmental Prediction (NCEP; Hamill & Colucci, 1997, 1998; Stensrud et al., 1999; Du and Tracton, 2001, Wandishin et al., 2001) during the Storm and Mesoscale Ensemble Experiment (SAMEX; Hou et al., 2001). Such probabilistic predictions have also been generated over the Pacific Northwest coast (Grimit & Mass, 2002) and over the Northeast coast (Jones et al., 2007) of the United States.

The combination of multiple models and multiple analyses is part of the operational suite of NCEP (Du & Tracton, 2001) and the basic idea of the short-range EPS of Washington University (Grimit & Mass, 2002) and the Agencia Estatal de Meteorología (AEMET; García-Moya et al., 2011).

2. Atmosphere as a chaotic system

2.1. Lorenz and non-linearity

Two basic properties can dynamically characterize a chaotic system: the sensitivity to initial conditions and the topologically mixing. Sensitivity to initial conditions implies that infinitesimal changes in the system initial trajectory can lead to big changes in its final trajectory. The Lyapunov exponent (Lyapunov, 1992) gives a measure to this sensitivity to initial conditions as it quantifies the rate of separation of infinitesimally close trajectories. Generally it cannot be calculated analytically and one must use numerical techniques. In Krishnamurthy (1993) it is described how to calculate the Lyapunov exponents of a simple system. The meaning of topological mixing is that the temporal evolution of meteorological quantities in any given region of its phase space will eventually overlap with those of any other given region. This second property is necessary to distinguish between simple unstable systems and chaotic systems.

The classical example provided by Lorenz (1963) is instructive. For this reason we use it in this section, to show briefly some concepts of Chaos Theory. It comes from a simplified model of fluid convection. It consists of a dynamical system with only three degrees of freedom, but it exhibits most of the properties of other more complex chaotic systems. It is forced and dissipative (in contrast to Hamiltonian systems which conserve total energy), non-linear (as its equations contain products of dependent variables) and autonomous (all the coefficients are time independent). The Lorenz (1963) equations are:

$$\frac{dx}{dt} = \sigma(y - x)$$

$$\frac{dy}{dt} = rx - y - xz \tag{1}$$

$$\frac{dz}{dt} = xy - bz$$

where, in this simplified model, $x(t)$ is proportional to the intensity of convection, $y(t)$ proportional to the maximum temperature difference between up and downward moving fluid portions and $z(t)$ is proportional to the stratification change due to convection. All variables are dimensionless, including time. The solution $\{x(t), y(t), z(t)\}$ is unique provided that the initial conditions $\{x_0, y_0, z_0\}$ are given at time $t = 0$. This means that the system is theoretically deterministic (given a perfect representation of the initial values or the dependent variables and a perfect integration of the non-linear system). The parameters $\{\sigma, r, b\}$ are constant within the time integration and different values provide different solutions thus creating a family of solutions of the dynamical system. Lorenz (1963) chose the values $\sigma = 10$, $r = 28$ and $b = 8/3$ which led to a chaotic solution of the system that is sensitive to small changes in the initial conditions and topological mixing. The dimension of the phase space is equal to the number of dependent variables (three in this case) whereas the dimension of the subspace reached by a given solution can be smaller as is the case of the Lorenz system. This behaviour can be demonstrated from the divergence of the flow:

$$\frac{\partial \dot{x}}{\partial x} + \frac{\partial \dot{y}}{\partial y} + \frac{\partial \dot{z}}{\partial z} = -(\sigma + r + b) \tag{2}$$

Which means that an original volume in the phase space V contracts in time to $Ve^{-(\sigma+r+b)t}$. This behaviour is related to the existence of a bounded attracting set of zero volume with dimension smaller than the phase space. An attractor is a set towards which the dynamical system evolves over time. Geometrically, an attractor can be a point, a curve, a surface or even a complicated set with a fractal structure known as a strange attractor. A solution of the Lorenz equations has an initial transient portion and after that it may be settled on a strange attractor. Figure 1 shows exemplarily a numerical solution of the Lorenz system up to $t = 100$ from with initial conditions equal to $x_0 = 0$, $y_0 = 1$ and $z_0 = 0$ using a backward Euler scheme for the time stepping with $dt = 0.01$.

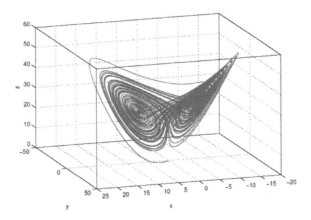

Figure 1. Numerical integration of the Lorenz (1963) system.

As is shown with more detail in next sections the difficulty of weather forecasting is due either to the sensitivity of the atmosphere evolution to small changes in the initial conditions related to the analysis error and to the sensitivity of the atmospheric differential equations to small differences in the numerical schemes used to find a numerical solution or model error. Figure 2 shows the evolution of the Lorenz system for two different but similar initial conditions. The solutions are very similar up to t = 25 approximately in this case and after that the differences become larger. After t = 30 the value of the variables x and y cannot be predicted although z remains more predictable. In general, the time range within which the system remains predictable, depends on the initial condition, and this characteristic is called the flow dependency of the predictability of the system.

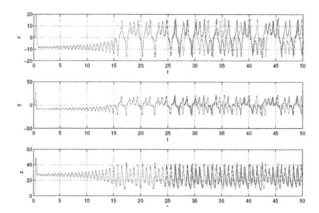

Figure 2. Numerical integration of the Lorenz (1963) system for two different and similar initial conditions. In green x_0 = 0, y_0 = 1 and z_0 = 0, in blue the initial conditions are x_0 = 0.001, y_0 = 1 and z_0 = 0.

The effect of model errors can be shown by changing slightly the constant parameters σ, r, b (Lorenz 1963). In a more complex model, this change would correspond, for example, to a change in the parameterization of the physical processes. Figure 3 shows the temporal evolution of the Lorenz system for two different sets of constant parameters. In this case, the predictability is loose after t = 20 for all the model variables.

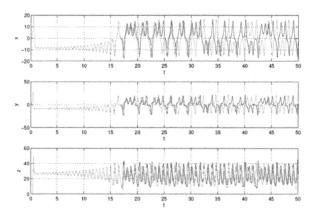

Figure 3. Numerical integration of the Lorenz (1963) system for two different and similar parameters. In green s = 10, r = 28 and b = 8/3, in blue a small value 0.001 is added to the parameters.

In light of these results (Lorenz, 1963), the question about the predictability of the atmosphere was raised for the first time, which has involved the efforts of the meteorological community to quantify it over several decades until today.

2.2. The predictability problem

A rigorous analysis of the chaotic properties of such a complex system as the atmosphere can be only achieved in simplified contexts. Atmosphere dynamics has been stated as chaotic and it is well established that there is an effective time barrier beyond which the detailed prediction of the weather may remain impossible (Lorenz, 1969). Predictability, or the degree to which a correct forecast can be made, depends on the spatial and temporal scales (from few hours at the mesoscale to few weeks at the planetary scale) and also on the variable (for instance, surface wind and temperature, precipitation or cloudiness).

Atmospheric chaos, uncertainty, predictability and instability are related concepts. Due to the approximate simulation of atmospheric processes, small errors in the initial conditions and model errors are the two main sources of uncertainties that limit the skill of a single deterministic forecast (Lorenz, 1963). Uncertainty limits the predictability, especially under unstable atmospheric conditions. The atmospheric instabilities related to low predictability conditions are the baroclinic instability at synoptic scales (Buizza & Palmer, 1995) and inertial and potential instabilities (e.g. deep convection) on the mesoscale, among others (Hohenegger & Schär, 2007; Zhang, 2005; Roebber & Reuter, 2002; Emanuel, 1979). This inherent limitation in predictability has led to the concept and development of ensemble prediction systems, which provide probabilistic forecasts to complement the traditional deterministic forecasts (Palmer et al., 1992).

3. Ensemble prediction systems

3.1. Uncertainty sources in numerical weather prediction

As indicated before, due to the chaotic nature of weather, there are several uncertainty or error sources in the Numerical Weather Prediction (NWP) framework that can grow and limit the predictability (Lorenz, 1963, 1969) of atmospheric flow. Forecast errors can arise due to inaccuracies in the initial condition atmospheric state estimates or due to imperfect data assimilation systems (Initial Conditions forecast error source), inadequacies of the NWP models themselves (Model Formulation forecast error source). Processes that take place at spatial scales that are shorter than the truncation scale of NWP models must be parameterized with sometimes inexact approximations thus giving us another source of forecast error (Parameterization forecast error source). One approach of NWP is to use Limited Area Models (LAMs) where the lateral conditions come from a global NWP models. This procedure is another source of forecast error (Lateral Boundary Conditions forecast error source). So far, the main error sources are: Initial Conditions (IC), Model Formulation, Parameterization and Lateral Boundary Conditions (LBC) error sources.

To the extent that these error sources project onto dynamical instabilities of the chaotic atmospheric system, such error will grow with time and evolve into spatial structures favoured by the atmospheric flow of the day. The inherent atmospheric predictability is thus state-dependen

To estimate these uncertainties or errors, i.e. the predictability, many operational and scientific centres produce ensemble forecasts (e.g. NCEP, ECMWF, etc.). The idea of using ensemble forecasts has been know for many years (Leith, 1974). Since the early 1990s, many centres generate ensemble forecasts. The methodology that is behind is to run multiple (ensemble) forecast integrations from slightly perturbed IC (IC forecast error source), using multiple models or perturbing model formulation (Model Formulation forecast error source). Adding stochastic physics parameterizations (Ehrendorfer, 1997; Palmer, 2001) or using multiple boundary conditions (Lateral Boundary Conditions forecast error source) among others techniques is described below. The discrete distribution of ensemble forecasts can be inter- preted as a forecast Probability Density Function (PDF). If an idealized forecast ensemble can be constructed that properly characterizes all sources of forecast errors, then the forecast PDFs would be reliable (see section 5) and skilful (sharper than the climatological PDF). No further information would be needed to make trustworthy forecast-error predictions, since a perfect PDF is a complete statement of the actual forecast uncertainty.

In practice, estimates of all the forecast-error sources mentioned above are inexact, leading to PDFs from real ensemble forecasts with substantial errors in both of the first two moments (mean and variance). These limitations are particularly pronounced for mesoscale prediction of near-surface weather variables, where large underdispersion results from insufficient ensemble size, inadequate parameterization of sub-grid scale processes, and incomplete or inaccurate knowledge of land surface boundary conditions (Eckel & Mass, 2005). Real ensemble forecast distributions, although generated using incomplete representations of weather forecast error sources, often represent a substantial portion of the true forecast uncertainty.

3.1.1. Initial conditions forecast error source

It is clear that the atmospheric state at a given time is not perfectly known; not only the inherent observational errors alone guarantee this, but also the sparse network of observations world- wide that sample the atmosphere only at limited intervals with inexact results. In addition network density and design can yield errors in regional averages (PaiMazumder & Mölders, 2009). Another contribution to IC forecast error source is the Data Assimilation (DA) system used. Every DA system is affected by the characteristic errors of both the observations incorporated in the analysis and of the short-range model forecast, which is typically used as a background or *first guess* field to be adjusted by new observations. IC errors, however small, are exacerbated by the chaotic dynamics of the atmosphere and consequently grow non- linearly with time.

3.1.2. Model formulation forecast error source

NWP model inadequacy is inevitable given to our inability to represent numerically the governing atmospheric physical laws in full. Contributions to this forecast-error source can be found in the model used which is, of course, a simplified scheme of what really happens in the atmosphere, dynamical formulation, different discretization methods, the numerical method employed to integrate the model and the different horizontal and vertical discretiza-tion resolutions used.

The model formulation forecast error in conjunction with another forecast error sources such as parameterizations has been recognized traditionally by operational forecasters in NWP centres. They usually select *a best model of the day* when producing their operational forecasts. This model selection tries to best handle the evolution of the atmosphere depending on the flow the general situation and the season of the year. The selection is driven by the subjective knowledge than some models behave better than others in some situations due to their formulation.

3.1.3. Parameterization forecast error source

There are several parameterized processes in NWP models: those which are taking place at smaller spatial scales than the truncation scale of the NWP model and are not resolved explicitly by the model as convection. Another one is introduced in a simplified way due to computer time limitations like radiation, and finally processes which are not taking into account in the NWP model dynamic part as microphysics in clouds. All theses processes are called sub-grid processes. It is assumed that sub-grid processes are in equilibrium with grid resolved states and so they can be represented statistically from them. A parameterization is the statistical method used when representing the sub-grid processes. Parameterizations are always imperfect representation of atmospheric processes so they always include inherent errors (Tribbia & Baumhefner, 1988; Palmer, 1997). NWP parameterizations have a time and space scale dependency. At small scales, forecast verification is primarily concerned with the locations and amounts of precipitation and other sensible weather parameters, which are often directly affected by the assumptions used to develop the model parameterization schemes for convection and other processes. Moreover, especially for the higher model resolutions, the implicit equilibrium assumption of sub-grid processes with model state could break down being another source of parameterization uncertainty.

3.1.4. Lateral boundary condition forecast error source

The LBC forecast error is only present in LAMs or regional models, which have as inputs lateral boundary values spatially and temporally interpolated from a coarser resolution grid-point or spectral model. So the coarser model errors are translated into LAMs as LBC error source. For instance, a possible configuration for a LAM EPS could include lateral boundary conditions from an ensemble of global forecasts.

3.2. Techniques used by global models

For many years operational forecasters, particularly medium range forecasters in meteoro-logical services, have had access to some forecast products coming from global NWP centres other than their own. They routinely compare forecasts from different centres to assess the confidence in the forecasts of their own models, and to determine possible alternative forecasts. This set of available forecasts is often called the *Poor Man's* (Ebert

2001) ensemble because its production is relatively cheap compared to the cost of developing and running a full EPS such as the ECMWF and NCEP ones. It is called *Ad hoc* ensembles by some other authors. Theses ensembles are cheap and easy to create, but they are not generated in a controlled and systematic approach. Not only are they not calibrated but also some ensemble members may be always quite more skilful than others. The hypothesis of equiprobability of the ensemble members is less guaranteed than others EPS which is a major drawback.

Hoffman and Kalnay (1983) proposed a *time-lagged* method or *Lagged Average Forecast* (LAF) method. The time-lagged method uses forecasts from lagged starting times as ensemble members. These members are easy to construct but they lack any scientific motivation. On the contrary, LAF perturbations are realistic short-term forecast errors. However, LAF ensemble forecasting has the disadvantage that most of the times earlier forecasts are considerably less skilful than later forecasts. This drawback can be partly resolved by either using different weights for different members of the ensemble or by scaling back the larger errors to a reasonable size. This procedure is the basis of the Scaled Lagged Average Forecast (SLAF) technique (Ebisuzaki & Kalnay, 1991).

The *multi-model SuperEnsemble* technique (Krishnamurti et al., 1999) is a powerful method to construct EPS. Several different models outputs are put together with appropriate weights to get a combined estimation of weather parameters. Weights are calculated by square minimization in a period that is called *training period*.

A better solution is to sample the different error sources that were indicated before. Depending on the sampling technique we obtain different methods: a Monte Carlo approach as proposed by Leith (1974), Hollingsworth (1980) and Mullen and Baumhefner (1989) among others. In general, the technique consists of sampling all sources of forecast error, by adding or perturbing any input variable (analysis, initial conditions, boundary conditions etc.) and whatsoever meteorological parameter that is not perfectly known. These perturbations can be generated in different ways. The main limitation of the Monte Carlo approach is the need to perform a high number of perturbations in order to have a proper description of the initial uncertainty, which is usually far from the available computational resources. This limitation leads to reduced sampling by just sampling the leading sources of forecast error due to the complexity and high dimensionality of the system. Reduced sampling identifies active components that will dominate forecast error growth.

IC forecast error source have a dominant effect. To sample it several techniques have been available. One of them is the initial perturbations method, which consists of adding small perturbations to the initial analysis, such as NCEP's *breed mode* method (Toth & Kalnay, 1993, 1997; Tracton & Kalnay, 1993). The breed mode method is based on the idea that the analysis created by the data assimilation scheme used will accumulate growing errors. As it can be seen in Figure 4 breeding vectors give a sampling of the growing analysis error: a random perturbation is added to the analysis, evolved in time by integrating the forecast model, rescaled and reintroduced as a perturbation. After several cycles only the fastest growing errors remain.

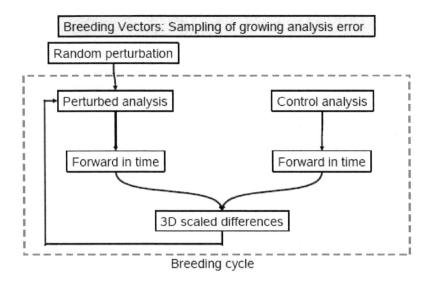

Figure 4. Breeding technique strategy

An alternative to the breed mode is the ECMWF's *singular vector* method (Palmer et al., 1992; Molteni et al 1996) which tries to identify the dynamically most unstable regions of the atmosphere by calculating where small initial uncertainties would affect a 48 hour forecast most rapidly. It needs an adjoint model. Singular vectors give a sampling of the perturbations that produce the fastest linear growth in the future. There are only a relative small number of directions in the phase-space of the atmospheric system along which the most important processes occur. Maximum growth is measured in terms of total energy. The adjoint of the tangent forward propagator with respect to the total norm is defined, and the singular vectors (the fastest growing perturbations) are computed by solving an eigenvalue problem. Singular vector method is schematically described in Figure 5.

In addition to the breeding and singular vector methods there are *Ensemble Transform Kalman Filter* technique (ETKF; Bishop et al., 2001; Wang & Bishop, 2003) and *Ensemble Data Assimilation* (EDA; Houtekamer, 1996; Buizza, 2008). ETKF is similar to the breeding method except that the rescaling factor is replaced by a transformation matrix. It produces an improved ensemble dispersion growth. It is used at the UK Meteorological Office. In EDA, an ensemble of assimilations is created from different analyses which have been generated by randomly perturbing the observations in a manner consistent with observation error statistics.

Model forecast error source, i.e. model formulation and parameterization error sources together, is another component to take into account. To represent model uncertainty several approaches have been used: the *multi-model* approach (e.g. DEMETER; ENSEMBLES; TIGGE; Krishnamurti et al, 1999), *multi-parameterizations* or *multi-physics* approach (Houtekamer, 1996), *stochastic parameterizations* (Buizza et al., 1999; Lin & Neelin, 2002), *multi-parameter*

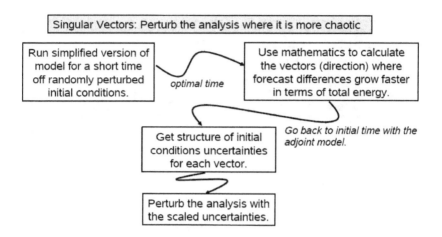

Figure 5. Singular vector technique strategy

approach (Murphy et al., 2004), and *Stochastic-Kinetic Energy-Backscatter* approach (Shutts & Palmer, 2004; Shutts, 2005).

So, in order to sample model forecast error an ensemble forecasts are produced by using different numerical models (multi-model approach). The multi-model approach implies equiprobable members which is not always the case. An alternative method for sampling model forecast errors is using different physical packages (multi-physics approach). Another approach is the *stochastic parameterization* approach applied at ECMWF (Buizza et al., 1999). It is based upon applying stochastically perturbing the total parameterized tendencies with a multiplicative noise. The multi-parameter approach tries to take into account the significant uncertainty in some parameters in NWP models, for instance, by using different values in each ensemble member.

Finally, the *Stochastic-Kinetic Energy-Backscatter* approach addresses a missing physical process, the upscale energy cascade energy from the grid scale to synoptic scales lost due to the excessive dissipation energy in NWP models.

3.3. Techniques used by limited area models

Not only global models can be used in building EPS, but also Limited Area Models (LAM) can be used to create LAM EPS, normally used for the short range. Error sources in LAM EPS are the same as in global EPS, but LAM models require lateral boundary conditions that update the weather situation regularly throughout the integration. These lateral boundary conditions introduce a main source of uncertainty in LAM ensembles. Both LBCs and ICs give their contribution to the spread and skill of the system (Clark et al., 2009). All the techniques discussed so far can be applied to generate LAM EPS. A very popular generating technique is the downscaling of global ensemble forecasts. This technique consists of using the selected global ensemble members (chosen by clustering) as initial and boundary conditions for a limited area ensemble system. The difficulty is that the perturbations generated from the global EPS are usually effective only on the medium range and large scales. Therefore they are not likely optimal for short range ensemble forecasts. Another technique for sampling lateral boundaries forecast error source is *multi-boundary* technique. In the multi-boundary technique, several different global models supply the lateral boundary conditions needed by the LAM model. One example of the use of the multi-boundary technique is the AEMET Short Range Ensemble Prediction System (AEMET-SREPS; García-Moya et al., 2011). AEMET-SREPS uses the multi-boundary method in addition to the multi-model method. It is built by using a set of LAMs and a set of deterministic global models that supply the initial and boundary conditions. The system is focused on short-range forecast and has been developed to help in the forecast of extreme weather events (gales, heavy precipitation and snow storm) and provides forecasts with good reliability, resolution and discrimination consistently with the analysis in the large-scale flow.

4. Uncertainty representation and weather forecasting products

4.1. Uncertainty representation

In statistics, uncertainty is represented by means of the Probability Distribution Function (PDF). Let us consider a random variable x that we do not know, a priori, anything about its nature. The question is whether we can infer something about it. Let us take n different values of x that belong to the same population. When we construct the histogram of these values, we obtain an approximation of its PDF. As an example, we could think of x as the mean monthly temperature of April at a surface observation station. Then the population would be *the mean monthly temperatures of April at that station*. If we restrict us to only the period 1981-2010, then the $n=30$ values of x would form the sample space.

The PDF gives us information about the behaviour of the random variable x. For example, let us take the normal or Gaussian PDF of which the analytical formula is:

$$PDF(x) = \frac{1}{\sqrt{2\pi\sigma^2}} e^{-\frac{(x-\mu)^2}{2\sigma^2}} \qquad (3)$$

Here σ is the standard deviation and μ is the mean. Figure 6 shows this distribution graphically. Now we can infer something about the nature of the variable x. From Figure 6 we can say that there is a value μ around which all the random variables are distributed symmetrically. Likewise, σ is a measure of the standard deviation of x from its mean. We can think of σ as a mean error (or uncertainty) we would have if we approximated any possible value of x by μ. In resume, the PDF gives us a depiction of all the possible values of x and their associated probabilities of occurrence. This procedure results in an explicit and quantitative way of representing the uncertainty of a random variable.

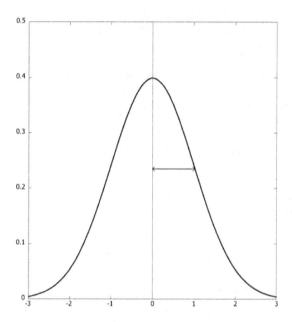

Figure 6. Gaussian PDF with μ = 0 and σ = 1

In the case of a Numerical Weather Prediction (NWP) system there are about 10^8 different random variables x, each one corresponding to each degree of freedom of the model. This fact makes it computationally unfeasible to integrate the Liouville equation (sometimes referred to as Fokker-Plank equation when random processes are included to account for, for example, model error) that describes the time evolution of a PDF. A practical way to resolve this problem is to use an EPS. An ensemble prediction tries to estimate the uncertainty of the forecast by discretizing the forecast PDF for each model parameter at each grid point in N values corresponding to the N ensemble members. As an example, Figure 7 presents the PDF of a 60 hours two metres (2m) temperature forecast of the AEMET-SREPS for the grid point closest to Sevilla, Spain. This ensemble has 25 members, but in this case, there was one that did not integrate properly, so $N=24$. It is easy to see that the more members the ensemble has, the higher is the resolution of the PDF.

In ensemble prediction, a simplified way of representing the uncertainty of the forecast is the *spread* (Toth & Kalnay, 1997); the standard deviation σ (4) of the PDF quantifies how much the ensemble members deviate, on average, from the mean, and it is often used as a measure of the spread (Wilks, 2006):

$$\sigma = \sqrt{\frac{1}{N}\sum_{i=1}^{N}(f_i - em)^2} \tag{4}$$

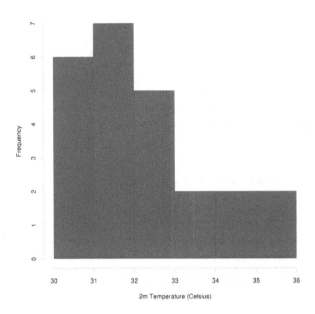

Figure 7. Histogram of 2m Temperature. AEMET-SREPS H+60 predicted values at the grid point closest to Seville. Valid time: 30th June 2011, 12 UTC.

Where f_i is the forecasted variable by member i, and *em* is the ensemble mean, e.g., the mean of the N forecasts. This parameter can be calculated on each grid point. In the case of the 2m temperature forecast (Figure 7) *em* and the spread are 32.3 and 1.5 °C, respectively. The latter can be interpreted as an estimate of the error of the deterministic forecast so that the higher the spread, the more uncertain is the forecast. Other measures of spread, more robust or resistant, can be alternatively used, e.g. the interquartile range (Wilks, 2006).

4.2. Raw products

Ensembles are composed of members that are deterministic predictions, and allow providing individual deterministic information (García-Moya et al, 2011). This information can either help the traditional staff to understand ensembles, and can provide support in the probabilistic interpretation. Far beyond this, ensembles provide their most useful information when we look at them as intrinsically probabilistic prediction systems. In this context, most of the ensemble outputs reflect this probabilistic nature. Probability is a rich and reasonable model to describe and understand many aspects of the physical world, but the interpretation of ensemble outputs must be learned and used carefully beyond the straightforward interpretation, because (especially for *deterministic forecasters*) some interpretations can be in conflict with *common sense*. Given a grid point with N forecast values x'_i (for an ensemble with N members), we call raw products when only these N values are used straightforward. Three basic examples of raw products exist.

4.2.1. Stamps

The deterministic outputs for all the ensemble members can be plotted as usual meteorological charts (see for instance Figure 8 with MSLP and T850 for the ECMWF EPS). These usual *postage stamp charts* comprise the control forecast (if there is any, top left), the perturbed members (below) and the corresponding high resolution deterministic forecast (if exists, beside the control). The difficulty is that the forecaster would have to deal with an amount of information: 51 scenarios in addition to the high resolution deterministic forecast.

4.2.2. Plumes

In a given location, we can provide N forecast values for that place (either by bi-linear interpolation or some other fine process which could account for height variability). Moreover, we can plot the evolution with forecast step for all the N members, i.e. we would plot N curves. Like on the stamp charts, the control forecast can be highlighted and the high-resolution deterministic forecast can also be plotted (e.g., Figure 9). The difficulty is similar to that one of the stamps namely the necessity to deal with such an amount of information. For a specific location and parameter, plumes can help the forecast guidance and, in fact, often provide information about the uncertainty and general trends.

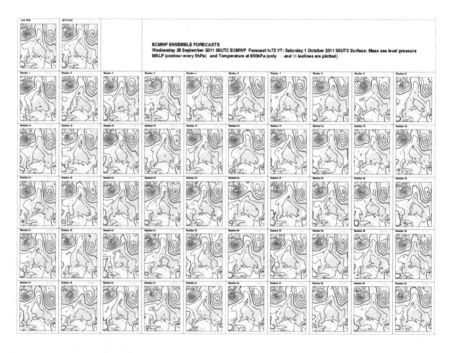

Figure 8. Postage stamps charts of MSLP and 850 hPa temperature T+108 forecasts (see text). Courtesy of ECMWF (2011).

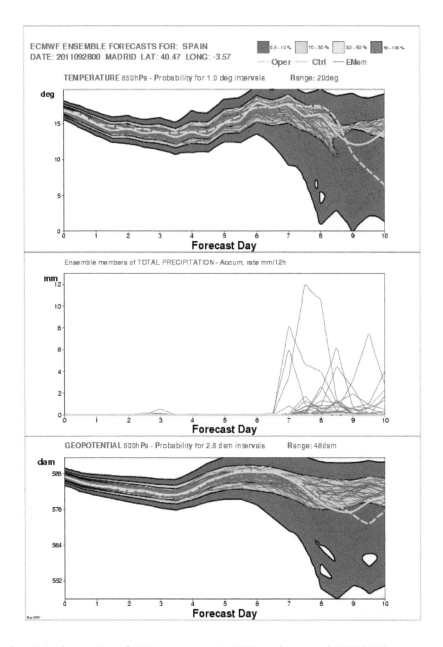

Figure 9. days forecast plumes of 850 hPa temperature at Madrid (see text). Courtesy of ECMWF (2011)

4.2.3. Spaghetti

As a third example, we show the spaghetti charts. For a given (often dynamical) field, it is rather impossible to overlay N member charts. But picking a selected isoline of interest, we can plot one line per member and, thus, the whole plot would contain N lines. Typically, the control is highlighted, and a higher resolution deterministic forecast can be included (e.g. Figure 10). As plumes, this kind of plot can help the forecaster to provide information about the uncertainty.

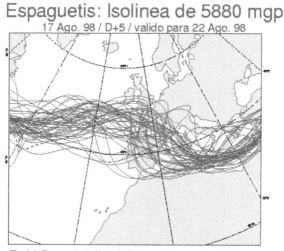

Espaguetis: Isolinea de 5880 mgp
17 Ago. 98 / D+5 / valido para 22 Ago. 98

Fig. 3.1: Espaguetis; isolinea de 5880 mgp del campo Z - 500 hPa.

Figure 10. Spaghetti chart of geopotential height (5880-gpm- isoline) for the T+120 forecast at 500 hPa (see text). Courtesy of ECMWF (2011).

All of these raw outputs share the same shortcoming: the inherent difficulty in the forecast guidance for handling the huge amount of information they provide. This issue is addressed using derived probabilistic outputs that compact this information naturally.

4.3. Derived probabilistic products

Probabilistic outputs are derived computationally from the PDF representation, which is assumed to be provided by the ensemble members. These products reflect the probabilistic nature of the ensemble, visually and conceptually. They provide explicit, quantitative and detailed information about uncertainty, and this fact is a real breaking point with respect to deterministic model products. They address the issue of providing compact information in this natural way. Ensembles nowadays can provide ideal complementary information to a higher resolution deterministic model.

For a given grid point, there are N forecast values x'_i for an ensemble with N members. Without further information about the skill of the members, we assume Laplace principle of equal probability, dealing in this case with a discrete PDF. An estimate of the forecast probability p of exceeding a threshold t is given by the well-know formula where the indicator $I(x'_i)$ is usually defined as $I(x'_i)=1$ if $x'_i > t$, $I(x'_i)=0$ otherwise (Ferro 2007b):

$$p = \frac{1}{N}\sum_{i=1}^{N}I(x_i') = \left\{\frac{0}{N}, \dots, \frac{N}{N}\right\} \tag{5}$$

The corresponding inverse is the computation of percentiles, i.e., for a given probability p, what is the actual forecast value x' for which $p = P(x')$. By adding further information, we can improve the PDF (e.g. by calibration) and the computation would be different. By taking this simple discrete model, we can compute the probabilities of exceeding thresholds, the percentiles for given probabilities, the summary statistics (e.g. mean and standard deviation), etc.

4.3.1. Ensemble mean and spread charts

The ensemble mean (the arithmetic mean of all the ensemble members) is not always a feasible meteorological situation because it is obtained as a result of a statistical operation, not from a numerical model (Buizza and Palmer, 1997). So, it is strongly discouraging in forecast guidance to use the ensemble mean without special care, if at all (García-Moya et al, 2011). However, the ensemble mean is often plotted in charts together with the standard deviation (the latter as a measure of spread), to help with the understanding of the atmospheric flow (e.g. Figure 11). The standard deviation is sometimes normalized.

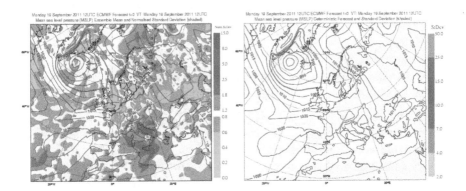

Figure 11. MSLP T+00 ensemble mean (contour) and normalized standard deviation (colours) (see text). Courtesy of ECMWF (2011)

4.3.2. Probability maps and percentile maps

Given a binary event (e.g. precipitation forecast > 5 mm/6h) we can plot the spatial distribution of the forecast probabilities that the EPS provides (see Figure 12). Similar plots can be made for the percentiles. These maps provide the forecasters with useful guidance by showing them where it is more probable for an event of interest to occur (e.g. representative precipitation that exceeds 5 mm/6h).

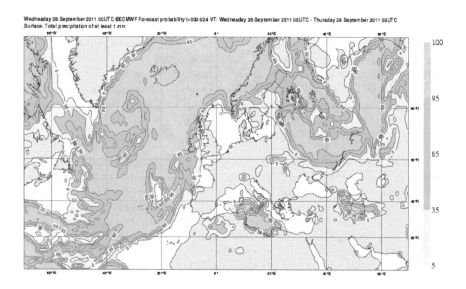

Wednesday 28 September 2011 00UTC ©ECMWF Forecast probability t+000-024 VT: Wednesday 28 September 2011 00UTC - Thursday 29 September 2011 00UTC
Surface: Total precipitation of at least 1 mm

Figure 12. Probability of having accumulated precipitation greater than 1 mm in the interval T+0 to T+24 (see text). Courtesy of ECMWF (2011).

4.3.3. EPS-grams

Box-plots (Wilks, 2006) and similar graphs give a quick, visual and simple representation of a distribution of numbers, a discrete PDF. Extending this idea by including the evolution with forecast time of the main weather parameters at a given location, we obtain plots that are *meteogram based* and often called *EPS-grams*. The building brick is the box-plot: it displays the median, minimum, maximum, percentiles 25 and 75 and sometimes also percentiles 10 and 90. Box-plots are displayed for a series of forecast steps. This procedure is typically applied for the more sensitive parameters to forecast e.g. cloud cover, precipitation, ten metres (10m) wind speed and 2m temperature (e.g. Figure 13). Special care must be taken with EPS-grams interpretation (Persson & Grazzini, 2005) by comparing the location point and the nearest grid-points: distance, land/sea contrast and height must be checked in order to properly use the information that EPS-grams provide. Anyway, the EPS-grams are the most popular and probably useful plots to forecast the weather in a location by taking into account the uncertainties.

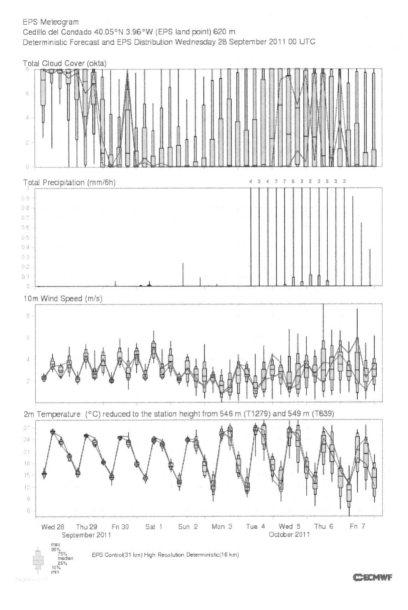

Figure 13. days EPS-gram at Cedillo del Condado (Spain) (see text). Courtesy of ECMWF (2011)

4.3.4. Clustering

A natural way of examining the number of atmospheric scenarios that EPS provide (e.g. stamps) is by similarity: one can gather scenarios in groups that could lead to similar weather conditions (Ferranti, 2010). This process can be done by *eye-ball* or can be carried out computationally by using a clustering algorithm (Ferranti, 2010) that fits the corresponding meteorological requirements. This procedure is often expensive and requires an extra task of defining the similarity criterion, but forecast guidance can improve substantially with the use of clusters. Examples of algorithms used are the Ward algorithm (Ferranti, 2010) and the tubing; both have been used operationally at AEMET (Figure 14). Here clusters have proved to be a very useful guidance in medium range forecasts by summarizing the more important and distinct scenarios (Ferranti, 2010).

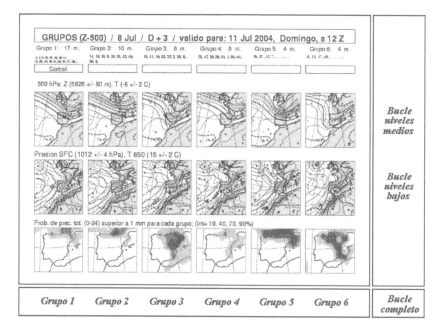

Figure 14. day forecast clusters of 500 hPa height and surface pressure (see text). Courtesy of AEMET (2011)

4.3.5. Extreme forecast index

Extreme events are not always severe, but severe events are often extremes. An index of *extreme forecast* can be computed using the model climatology as a reference, rather than the climatology of observations (Lalaurette, 2003). When observations are used, the forecast is not really prone to fall in the tail of the climatological distribution, and this fact is addressed by using the model climatology. The Extreme Forecast Index (EFI; Lalaurette, 2003) is a quantitative measure of how extreme is the EPS forecast when compared with the model climatology. The EFI can be plotted in a chart (Figure 15), and this chart is especially useful for weather parameters. Thus the EFI is used by forecasters as an early warning tool to highlight where severe events could happen.

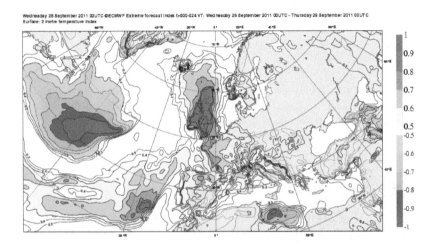

Figure 15. Extreme Forecast Index for range T+0 to T+24 of 2m temperature forecast (see text). Courtesy of ECMWF (2011).

4.4. Interpretation for weather forecasting

As an example for the application of probabilistic forecasting we present a real case of extreme winds that is fully described in Escribà et al. (2010). This section is not intended to be a detailed manual of the utilization of probabilistic products in operational forecasting. More extend and concise information can be found for example at www.ecmwf.int.

At 00 UTC on 24 January 2009 an explosive cyclogenesis in the Gulf of Vizcaya reached its maximum intensity with an observed surface pressures less than 970 hPa on its center. During the cyclone's path through the south of France strong westerly and north-westerly winds occurred over the Iberian Peninsula (> 150 km/h). These winds led to eight casualties in Catalunya, the north-east region of Spain.

In Figure 16 are represented three probabilistic forecasting products, the ensemble mean, the ensemble spread and the probability of having wind speeds greater than 15 m/s (54 km/h). The weather parameter analyzed is 10m wind speed. These fields correspond to the H+60 predic-tion of the AEMET-SREPS initialized at 00 UTC on 22 January 2009. The ECMWF reanalysis is also shown as verification.

Even though wind speed values plotted in the maps are not extreme, they do not exceed 20 m/s or 72 km/h, this fact has to be taken carefully because we are representing mean values of wind at a forecast time, i.e. a mean over a time interval equal to the last forecast time step of the forecasting model (which in this case is around 5 minutes). As a first approximation we can estimate the wind gust (maximum wind) as twice the value of the mean wind (this factor can be roughly obtained comparing temporal series of mean wind and wind gusts from observation ground stations). In this case, such an approximation would give us extreme winds of about 150 km/h, similar to those observed.

The ensemble mean (Figure 16) can be thought as a skilful deterministic forecast that comes from the ensemble. When we compare it with the verification we can highlight there is a good agreement in the overall patterns. Looking in more detail we can select three zones where there is more discrepancy: a.) south of Catalunya (yellow ellipse), b.) Aragon and Valencia (white ellipse) and c.) south-east of France (green ellipse). It is especially interesting to analyze zone a.), where the casualties occurred. The question is whether the ensemble can estimate in some way the error in the prediction; the spread is expected to give information on this.

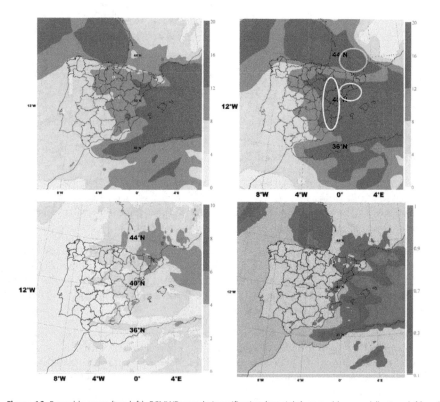

Figure 16. Ensemble mean (top left), ECMWF reanalysis verification (top right), ensemble spread (bottom left) and probability forecast of S10m > 15 m/s (bottom right). The probability field is the only one that is not in m/s. Lead time prediction is H+60 and verification time is 12 UTC of 24 January 2009.

The ensemble spread gives us the areas of more uncertainty in the prediction and is a measure of it. For zones a.) and c.) it has values around 5 m/s, which are considerable. In the case of a.), the spread estimates properly the prediction error, giving valuable information. When making the forecast, we would say that it is possible to have wind speeds greater than 16 m/s. This is not the case in zone c.), where the spread is not enough to explain the discrepancy between the ensemble mean prediction and the verification. In this case, the probabilistic forecast is badly displaced to the east. In zone b.), a lower spread (around 3 m/s) has also the ability to describe the prediction error.

Finally, the probability forecast enforces the general forecast by determining the areas of maximum confidence of occurrence and quantifying this confidence in a number. In this sense, we can say that the probability of having mean wind speed greater than 15 m/s (54 km/h) or wind gusts of more than 100 km/h in zone a.) is between 30% and 70%, which is more than the term *possible*.

5. Ensemble forecast verification

NWP models must be compared with a good representation of the observed atmosphere. This process is often called *forecast verification*, and raises a number of concepts and issues. With verification, we assess the quality and value of forecasts (Murphy, 1993), by using metrics or measures often called scores. By providing detailed information about forecast performance, verification can help in model improvement (developers) and forecast guidance (forecasters). Comprehensive descriptions of standard verification methods can be found in Wilks, 2006 and in Jolliffe & Stephenson, 2003), whereas in Candille & Talagrand, 2005 and in Stensrud and Yussouf, 2007 there is a thorough study of probabilistic forecasts, including ensemble forecasts.

Different frameworks are available for verification. Observations (ob) and forecasts (fc) can be compared, either *whole set to whole set* or *fc to ob* by using their statistical summary properties (measures-oriented approach), as distributions (distributions-oriented), as features (features-oriented), etc. In any case, to compare observations and models is like comparing apples and oranges: they are often different kinds of atmospheric representations, and we need to transform one or both of them into *comparable* representations. This step involves non-trivial issues like interpolation, representativeness, correlation, noise, etc.

An ordinary example of the difficult issue in comparing apples and oranges is the performance assessment of quantitative precipitation forecasts (QPF) from a deterministic model. European meteorological offices provide to the ECMWF 24-hour accumulated precipitation reports from their high-density rain gauges networks. Forecast values are regularly spaced, while observations are not. One comparison method is to interpolate forecast values to observation points (Rodwell, 2010). Special care should be taken with the impact of spatial density of observations and the potential lack of statistical consistency due to spatial dependence between close ones. To address these issues, one can compute an observed quantitative precipitation estimate (QPE) using a simple up-scaling technique (Ghelli & Lalaurette, 2000; Cherubini et al., 2002) whereby stations are assigned to model grid-boxes and then *averaged* to produce one value to

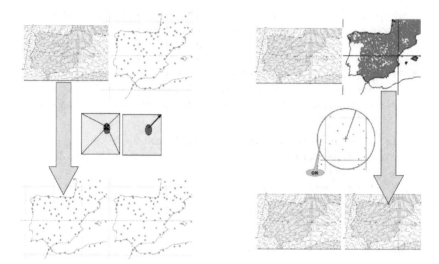

Figure 17. Interpolation to point (left) versus up-scaling (right)

be assigned to the corresponding grid-point. Such a grid value can be for instance a weighted average, and can be compared to the model precipitation forecast, which is also relative to the grid-box areal average, as they both refer to the same spatial scales. Whether to choose interpolation or up-scaling methods it depends on the case (see Figure 17).

Moreover, in the comparison of the performance of QPF from two different forecasting models, further issues arise. The grid spacing of the two models might be different. If observations are gridded to the finer resolution, then the coarser model might be penalized. On the other hand, if observations are gridded to the coarser resolution, the comparison can be fair but the higher resolution model is not given a chance. How to compare the way in which both models represent structures at their own scale is a non-trivial issue. PaiMazumder & Mölders (2009) assessed the impact of network density and design on regional averages using real sites and model simulations over Russia. They find that generally, the real networks underestimate regional averages of sea level pressure, wind speed, and precipitation while overestimate 2m temperature, downward shortwave radiation and soil temperature.

5.1. A first requirement: Deterministic performance of ensemble members

The assessment of the deterministic quality of the ensemble members is a first requirement in the development of an EPS. When the quality of the ensemble members is similar, then any member can be weighted equally in the computation of a probabilistic forecasts, i.e. they are assumed to be equiprobable. Once provided this individual quality, then some other properties can be considered (see below). In addition, the ensemble mean should show a better deterministic performance than any individual member in terms of Root Mean Square Error (*RMSE*) (Leith, 1974; Murphy, 1988; Whitaker & Loughe, 1998; Ziehmann, 2000).

As a visual representation, either time-series or evolutions with forecast step for bias and *RMSE* are usually depicted for synoptic parameters (e.g. Z500) for each member and also for the ensemble mean. As an example, Figure 18 shows *BIAS* and *RMSE* evolution with forecast length for the different ensemble members and the ensemble mean.

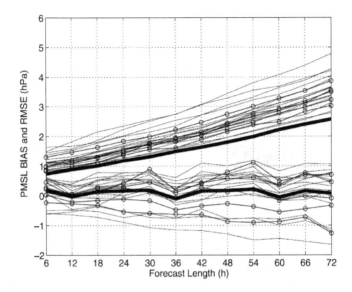

Figure 18. Evolutions with forecast length of mean sea level pressure (*MSLP*) *BIAS* (bottom group) and *RMSE* (upper group) computed for each member (thin lines) and for the ensemble mean (thick line). 'Normal operating' members are highlighted (circles) (García-Moya et al., 2011).

5.2. Large scale flow: Statistical consistency with the observations/analysis

As a probabilistic forecast, an EPS must be statistically consistent with the observations in the large scale flow given the EPS domain is large enough. At this scale, the model analyses of upper-air dynamical fields (e.g. 500 hPa geopotential height, Z500) can be used for comparison, by providing a larger sample and covering the whole integration domain, and so by giving no priority to land areas where the density of observations is higher. Verification against SYNOP/TEMP observations is expected to give worse but qualitatively similar results. This statistical consistency can be assessed in several ways; two methods are shown here: the rank histogram and the spread-error diagram.

On each grid-point, either the analysis or each of the ensemble member values are assumed to be independent realizations of the same atmospheric process and hence equally probable. Here, the rank of the analysis is an integer number according to the position of the analysis value in the sorted list of forecast values. The rank histogram (Anderson, 1996; Hamill & Colucci, 1997, 1998; Hamill, 2000; Candille & Talagrand, 2005) can be used to check if the

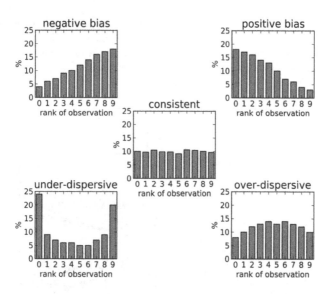

Figure 19. Examples of rank histograms (see text).

verifying observation is statistically indistinguishable from the set of forecast values (*reliable*). Such a system must show an approximately flat-shaped rank histogram (Figure 19 middle). Some outliers ("U" shape, Figure 19 bottom left) would indicate sub-dispersion that are typical in current EPS operational systems, while overdispersion would correspond to the opposite (inverted "U" shape, Figure 19 bottom right). The bias would produce rank histograms with positive (negative) or negative (positive) slope (bias) (see Figure 19 top left and top right, respectively).

Furthermore, the ensemble spread (often measured by the standard deviation with respect to the ensemble mean or the control forecast if possible) and the error of the ensemble (measured by the *RMSE* with respect to the analysis for either the control forecast or the ensemble mean) should show a linear relationship and a similar growth rate with respect to forecast step (Buizza & Palmer, 1997; Whitaker & Loughe, 1998). An EPS is expected to sample the uncertainties of NWP models (ensemble *spread*), as well as to give explicit and quantitative information about the predictability of the atmosphere (represented by the ensemble error). According to this, an ensemble can be statistically consistent (*calibrated*) or, on the other hand, can be underdispersive (quite common in operational ensembles) or overdispersive (e.g. Figure 20).

5.3. Weather parameters: Binary events

For the performance assessment of weather parameters (e.g. precipitation, 2m temperature, 10m wind), with larger variability in space and time, the use of observations is encouraged, as they are not as smooth as upper-air field analyses. In the distributions-oriented framework (for a detailed description see Joliffe & Stephenson, 2003; Wilks, 2006), the performance of an

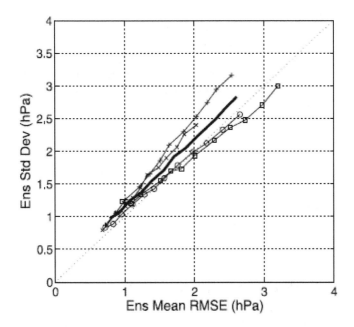

Figure 20. Spread-error diagrams showing five EPS: one of them consistent (solid black), two of them overdispersive (x crosses, + crosses) and two of them underdispersive (circles, squares).

EPS can be done measuring its response to a set of binary events (occurrence / non-occurrence, e.g. to exceed a threshold). The EPS behavior in this context is described by different properties: reliability, resolution, sharpness and discrimination. Brier (Skill) Score together with *ROC* curves provide a framework to give measures to theses properties. Moreover, the benefits of using a forecasting system can be shown with the so-called relative value, a quantity that depends on the forecast user's cost/loss ratio. In this framework, the joint distribution of forecasts and observations gives complete support for the computation of scores.

5.3.1. Formal framework

By considering a binary event X of the parameter x exceeding a threshold t ($\{X: x>t\}$ e.g. rain over 5 mm), we compare a forecast probability p (number of member values exceeding t, p = $\{0/N, 1/N,... N/N\}$) and the corresponding a posteriori observation probability p_0, which is usually taken as $p_0 = \{0,1\}$, depending on whether the event took place or not. However, if observational uncertainty was considered, then p_0 could take any value in the interval [0,1]. In this probability space, a natural extension of the *RMSE* is the Brier Score, defined as $BS = E$ $[(p - p_0)^2]$, where E[] is the expectation value over all forecast-observation pairs. BS is negatively oriented and $BS=0$ if and only if $p=p_0$ for any pair, while $BS=1$ indicates the worst possible forecast.

The *joint distribution* (Murphy, 1988) of forecasts and observations can be represented by two distributions that describe completely the system performance: $g(p)$ and $p'(p)$, where $g(p)$ is the forecast probability distribution (relative frequency of forecasts with probability p) and $p'(p)$ gives the conditional observation distribution (relative frequency of forecasts with probability p and for which the event did happen). The expectation values can be computed through a partition in probability space according to the possible forecast probability values, i.e., the number of members (Santos & Ghelli, 2011):

$$E\left[y\right] = E\left[E_p\left[y\right]\right] = \int_0^1 y\, g(p)\, dp \approx \sum_{p=0/N}^{p=N/N} y\, g(p) \tag{6}$$

The base rate $p_c = E[p'(p)] = E[E_p[p_0]]$ is the frequency of occurrence of the event. Using these two distributions, a decomposition of the BS can be done (Jolliffe & Stephenson, 2003; Candille & Talagrand, 2005):

$$BS = E\left[p - p_0^2\right] = E\left[(p - p'(p))^2\right] - E\left[(p'(p) - p_c)^2\right] + p_c(1 - p_c) = B_{rel} - B_{res} + B_{unc} \tag{7}$$

the components are meaningful: *Reliability* (B_{rel}) measures the straight correspondence between probabilistic forecasts p and conditional observation frequencies $p'(p)$, and can be improved by re-labeling of probability intervals (a process that should be called *re-labeling* calibration to avoid confusion). *Resolution* (B_{res}) gives a measure of variability of conditional observations $p'(p)$ around the base rate, and cannot be improved by re-labeling, thus it gives an upper bound for inherent skill. For a perfectly reliable system the reliability component vanishes ($p = p'(p)$ for all cases), and the resolution is equal to the *sharpness*, a measure of variability of the forecast probability distributions, or how often different forecast probabilities occur without taking into account the observations. The uncertainty component (B_{unc}) is the variance of the probabilistic observations p_0 and corresponds to the value of the BS using the sample climatology as forecast (i.e. issuing always a forecast probability $p = p_c$ a system is perfectly reliable $B_{rel} = 0$, and shows no resolution $B_{res} = 0$); it depends only on the observations and is usually taken as a reference for the Brier Skill Score (BSS), if special care is taken with the interpretation (Mason, 2004). The same decomposition can be applied to the BSS (Candille & Talagrand, 2005):

$$
\left.
\begin{aligned}
BSS_{rel} &= \frac{E\left[(p - p'(p))^2\right]}{p_c(1 - p_c)} \\
BSS_{res} &= 1 - \frac{E\left[(p'(p) - p_c)^2\right]}{p_c(1 - p_c)}
\end{aligned}
\right\}
\quad BSS = 1 - BSS_{rel} - BSSs
\tag{8}
$$

To give a summary of performance measures comprising the response to several thresholds, the Ranked Probability Score (and the corresponding skill score) can be used, either the discrete or the continuous version (Hersbach, 2000).

A complementary measure of ensemble performance is the *discrimination* or ability of a system to distinguish the ocurrence or non-ocurrence of a binary event X given the observations according to Signal Detection Theory (Kharin & Zwiers, 2003). The discrimination is related to the hit rate (H) and the false alarm rate (F) for a given base rate p_c (Candille & Talagrand, 2008):

$$\left\{ \begin{array}{l} H(s) = \dfrac{1}{p_c} \int\limits_s^\infty g(p)p'(p)dp \\[4mm] F(s) = \dfrac{1}{1-p_c} \int\limits_s^\infty g(p)(1-p'(p))dp \end{array} \right. \tag{9}$$

As a measure of discrimination, the area A under the Relative Operating Characteristics (ROC) curve (H versus F) is often used, with $A=0.5$ for the sample climatology (no skill) and $A=1$ for a perfect forecast (Santos & Ghelli, 2011). ROC Skill Area (RSA) can be used instead: if A is the area under the ROC curve, $RSA=2A-1$ gives values in the interval [-1,1], 1 for a perfect forecast, 0 for no skill and -1 for a potentially perfect forecast after calibration. Discrimination is related to resolution, but they do not measure exactly the same property and, especially if observational uncertainty is present, they can show different and indicative behaviour. While BSS is potentially insensitive to extreme events, RSA is not (Gutiérrez et al., 2004), whereas RSA can be insensitive to some kinds of forecast biases (Kharin & Zwiers, 2003).

Another interesting complementary property, beyond performance, is the Economic *Relative Value* (RV; Richardson, 2000). By crossing a *contingency table* (forecast yes/no by occurrence yes/no of the event) with an *expenses matrix* (preventive action yes/no by occurrence yes/no, that includes the cost C of the action and the loss L in case of occurrence), it can be computed the relative economic reduction (RV) of using the forecast comparing with the sample climatology. RV depends on the base rate p_c and also on C and L, i.e. it depends also on the user.

5.3.2. *Visual presentation*

The properties described above can be visualized in several ways. *Sharpness histogram*: $g(p)$ distribution is put in a histogram along probability intervals. A predicting system with good sharpness would issue forecast probabilities close to 0 and 1. Sharpness is often plotted as an inset on the attributes diagram. *Attributes diagram*: $p'(p)$ distribution is plotted on the Y axis against probability intervals p on the X axis. Several straight lines are plotted as reference: the diagonal (representing perfect reliability), the no-resolution line (corresponding to the sample climatology as forecast), and the no-skill line (forecasts with no skill w.r.t. the climatology, i.e. $B=B_{unc}$). Figure 21 (left) illustrates this. Some examples of forecasting systems are idealized in Figure 21 (right).

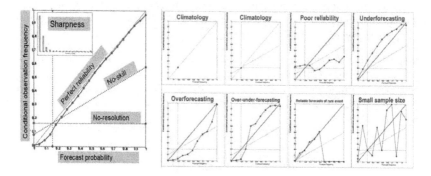

Figure 21. left) Attributes diagram for an almost perfectly reliable forecasting system, showing the sharpness histogram; (right) Attributes diagrams for idealized examples of forecasting systems.

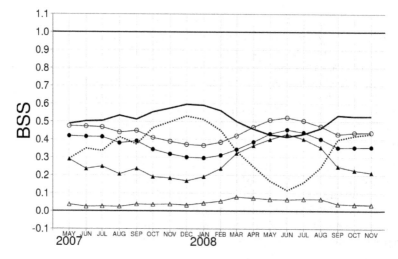

Figure 22. Time series for two different systems 24 h accumulated precipitation forecast (T+30 to T+54), of *BSS* (solid, dotted) and its components BSS_{rel} (triangles) and BSS_{res} (circles). Santos and Ghelli (2011)

BSS decomposition time series: *BSS* (and its components BSS_{rel} and BSS_{res}) time series are plotted in curves. As *BSS* is positively oriented (the larger the BSS, the better is the performance) and its components are not, special care must be taken (see Figure 22) (Santos and Ghelli, 2011). *ROC curves*: the hit rate (Y axis) is plotted against the false alarm rate (X axis) (see Figure 23 left). Here a deterministic forecast is compared to an EPS. *RV envelopes*: the *RV* can be plotted on the Y axis, the cost-loss ratio C/L on the X axis and provide one curve for a deterministic model. For an N members ensemble, N curves can be plotted (we can plot *RV* for any probability interval in the partition described above), or eventually, the envelope that covers all the

N curves. The user can decide on the C/L intervals for optimal use of the forecasting system. In this sense, an EPS can be also compared with a deterministic model (see Figure 23, right).

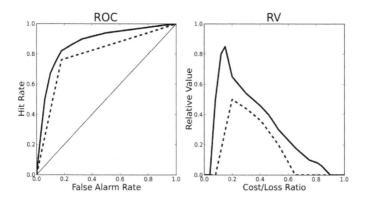

Figure 23. Comparison of deterministic (dash) and EPS (solid) precipitation forecast performance, using empirical *ROC* curves (left) and relative value envelopes (right).

5.4. Verification issues and prospective

Computational resources, object-oriented programming languages and data-base improvements give a boost to forecast verification. In the last decades, the forecast verification community has started to address some important issues that have an impact in the interpretation of verification scores and introduction of new conceptions. Either EPS specific or not, some of these issues are of large interest and hence are introduced here.

5.4.1. Pooling versus stratification

To compute statistically significant scores, samples must be large (many fc-ob pairs) and the corresponding significance tests should be applied (e.g. t-Student). On the other hand, mixing non-homogeneous sub-samples (e.g. different seasons) can lead to misleading performance information. In this context, the dimensionality problem (Murphy, 1988) can be a critical stumble in practice (Candille & Talagrand, 2008): computing completely consistent (from the strictly mathematical point of view) scores can turn out to be an infeasible task, and this issue often leads to a practical compromise: large samples are created without mixing different ones, according to possibilities. E.g.: the splitting of seasonal behavior that could be hidden in the overall average.

5.4.2. Flow-dependent verification

Flow-dependent sample stratification can improve other traditional ways of stratifying (e.g. seasonal), and nowadays can be tackled with clustering techniques (Ferranti & Corti, 2010

5.4.3. Sampling uncertainty

Given the dimensionality problem and pooling-stratification compromise, the available fc-ob pairs are, in practice, relatively small samples from the all possible realizations of the model and observing systems. Thus, the scores computed are only sample measures of the population quantities, and there is a sampling uncertainty related to this process (PaiMazumder & Mölders, 2009). Any verification report should include this uncertainty, with error bars, confidence intervals, etc. (see e.g. Efron & Tibshirani, 1997; Bradley et al., 2008).

5.4.4. Spatial scales of forecasts and observations

Spatial scales are a key point (see examples above). For example, the double penalty (Ebert & Gallus, 2009) is a well-know related issue. The relatively recent development of new methods that account for spatial patterns, e.g. CRA (Ebert & Gallus, 2009), MODE (Davis et al., 2009), SAL (Wernli et al., 2008) are still under research, but show promising results and can lead to a framework of diagnostic verification (closer to subjective verification in the sense that provides information that can better help model developers and weather forecasters). For a comprehensive overview, see (Gilleland et al., 2009).

5.4.5. Extreme and severe weather

Extreme and severe weather are often introduced together, but they are not the same. Extreme events are rare events, with low base rates and belong to the tail of the corresponding climatological distributions. Severe events are those that have an impact (human and material) on society. Severe weather verification must include extra information from outside the meteorological context, whereas verification of extreme events is still in an early stage (Ferro, 2007a; Casati et al., 2008), and some alternative scores are under research, e.g. the Extreme Dependency Score (EDS; Stephenson et al., 2008; Ghelli & Primo, 2009).

5.4.6. Observational uncertainty

In forecast verification it has been traditionally assumed that the observation error is negligible when compared with the forecast error. This assumption can be consistent for longer forecast ranges, when the forecast error is much larger than the observation uncertainty. Several studies have extended the verification problem to a more general framework, in which observations are described together with their uncertainty. They show sometimes a surprising result: traditional scores generally underestimate EPS performance (e.g., Saetra et al., 2004; Candille & Talagrand, 2008; Santos & Ghelli, 2011).

5.4.7. Ensemble size

Differences in ensemble size can have an impact on performance assessment (e.g. compare a 16 members EPS with a 51 members EPS). The difference in size would, in principle, give better performance to the larger EPS a fact that should be at least taken into account. This issue is addressed by various authors (e.g., Buizza & Palmer, 1998; Ferro, 2007b; Ferro et al., 2008).

6. Statistical post-processing

EPS evidence systematic errors like do the deterministic NWP models. Calibration is the process of correction of the ensemble PDF to adjust it to the actual (and unknown) forecast uncertainty. The main point of calibration techniques is to use the information of the former prediction's skill to correct the current probabilistic forecast.

Different methodologies have been proposed recently to build calibrated probabilistic forecasts from ensembles, including Bayesian Model Averaging (Raftery et al., 2005), Logistic Regression (Hamill et al., 2008) and Extended Logistic Regression (Wilks, 2009), Non-homogeneous Gaussian Regression (Gneiting et al., 2005b) and Ensemble Dressing (Roulston & Smith, 2003; Wang & Bishop, 2005).

6.1. Bayesian Model Averaging (BMA)

Bayesian Model Averaging (BMA) is a statistical post-processing method that generates calibrated and sharp predictive PDFs from EPS (Raftery et al., 2005). The BMA predictive PDF of a weather variable is a weighted average of PDFs centred on the individual bias-corrected forecasts. The weights reproduce the predictive skill of the ensemble member over a training period.

If the forecast errors are approximately Gaussian distributed such as surface temperature or sea level pressure, BMA can be applied (e.g. Raftery et al., 2005; Wilson et al., 2007). For non-Gaussian error distributions using a mixture of skewed PDFs allows to extend the BMA methodology to this kind of weather parameters; A combination of point mass at zero and a power-transformed gamma distribution, for instance, can be applied to quantitative precipitation (Sloughter et al., 2007) and a mixture of gamma distributions with different shapes and scale parameters can be used to improve wind speed probabilistic forecasts (Sloughter et al., 2010).

The BMA predictive PDF is a summation of weighted PDFs of each individual ensemble member (Leamer, 1978; Kass & Raftery, 1995; Hoeting et al., 1999):

$$PDF\left(y \mid f_1, \ldots, f_m; \theta_1, \ldots, \theta_m\right) = \sum_{i=1}^{m} w_i PDF_i\left(y_i \mid f_i, \theta_i\right) \qquad (10)$$

Where f_i is the ensemble member deterministic forecast, y represents the forecasted variable, and $_i$ are the characteristic parameters of the ith individual PDF from the ith ensemble member. Each of these individual PDFs associated to each ensemble member is weighted based on the ensemble member's relative performance during the training period. The weights w_i are probabilities, i.e. non-negative and add up to 1. The BMA weights w_i and the parameters $_i$ are estimated by maximum likelihood (Wilks, 2006) using the training data. This estimate cannot be done analytically so an expectation maximization (EM) iterative algorithm is used (Dempster et al., 1977; McLachlan & Krishnan, 1997).

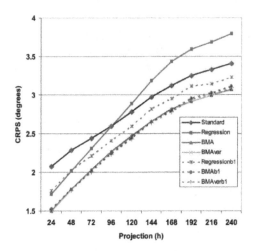

Figure 24. CRPS of surface temperature ensemble predictions at 21 ground stations over Canada. Curves correspond to raw ensemble and the six other ensembles corrected by: Linear regression, BMA, a variation of BMA, and the same three methods with the bias previously corrected. 40 days are used as training period., From Wilson et al. (2007). See their article for details. © American Meteorological Society. Reprinted with permission.

Figure 24 illustrates the potentiality of BMA as a method for ensemble calibration. Continuous ranked probability score (CRPS) performance score (Hersbach, 2000) is represented for various ensemble predictions of surface temperature. CRPS index is understood to be for a probabilistic forecast the equivalent of the mean absolute error for a deterministic forecast. The different curves correspond to the raw ensemble and to the six ensembles calibrated using different statistical techniques. It is straightforward to see that the ensembles corrected with BMA perform better than any of the others.

6.2. Logistic Regression (LR) and Extended LR

The logistic regression and extended logistic regression techniques are described in detail in Wilks (2009). Logistic Regression (LR) approximates the Cumulative Distribution Function (CDF) of the predicted parameter y by the following equation (Wilks, 2009):

$$CDF(q) = PDF(y \le q) = \frac{e^{f(x)}}{1 + e^{f(x)}} \tag{11}$$

Where q is a selected prediction threshold and:

$$f(x) = b_0 + b_1 x_1 + \cdots + b_n x_n \tag{12}$$

Being $\{x_1, ..., x_n\}$ the regression predictors and $\theta = \{b_0, b_1, ..., b_n\}$ the unknowns to be estimated during the training process. Equation (11) has a characteristic S shape with values bounded on the $0 < CDF(q) < 1$ interval. The name logistic comes to the fact that the regression equation is linear on the logistic scale:

$$\ln\left[\frac{CDF(q)}{1-CDF(q)}\right] = f(x) \tag{13}$$

Typical predictors for LR when calibrating ensemble predictions are ensemble mean, ensemble spread or a function of them (Hamill et al., 2008). As thresholds q, using the representative climatological quantiles of the meteorological parameter y ensures a statistical uniformity in the process of regression.

According to Wilks (2009) θ unknowns are generally estimated using maximum likelihood (Wilks, 2006), but other estimation techniques could give better performance, for example the minimization of the continuous ranked probability score (Hersbach, 2000).

By construction Equation (13) is fitted separately for every threshold q and this fact involves several problems. We consider for example the parameter precipitation and two thresholds, $q_1 = 2mm$ and $q_2 = 10mm$. After the training we have two different regression equations for each threshold, $f_1(x)$ and $f_2(x)$, which in general are not parallel. The non-parallelism of the functions implies that for some values of the predictors $\{x_1, ..., x_n\}$ these curves will cross leading to the unrealistic result of $CDF(q_1) > CDF(q_2)$. Another problem arises when we want to estimate the CDF of a threshold for which regressions have not been fitted. This process requires some kind of interpolation of CDFs which is not statistically coherent. Finally, the more equations are to be fitted the more unknowns have to be estimated.

To overcome these problems, Wilks (2009) proposed a new approach to Equation (11) that consists of including a function $g(q)$ in the exponent which increases with threshold q:

$$CDF(q) = PDF(y \le q) = \frac{e^{f(x)+g(q)}}{1+e^{f(x)+g(q)}} \tag{14}$$

Thus, a unique regression estimation for any value of q is needed, which implies the parallelism of the functions $f(x)$ for the different thresholds (the unknowns $\{b_0, b_1, ..., b_n\}$ are always the same). This approximation is known as Extended Logistic Regression (ELR).

It is important to point out as an advantage that LR (and ELR) has no statistical restriction to be used with non-Gaussian parameters such as precipitation or wind. At the same time this technique can be applied to ensembles whose members are non-distinguishable.

As an example of ELR, Schmeits and Kok (2010) calibrated ECMWF ensemble predictions of precipitation over an area that covers Netherlands (see the article for details). After studying the performances of different shapes for $g(q)$ and the predictors, they select :

$$g(q) = b_2\sqrt{q}$$
$$x_1 = \left(\sqrt{x}\right)_{ens}$$

(15)

Being x_1 the ensemble mean of the square root of the predicted precipitation. Then the equation to be regressed is:

$$f(x) + g(q) = b_0 + b_1\overline{\left(\sqrt{x}\right)}_{ens} + b_2\sqrt{q}$$

(16)

Figure 25 represents a reliability diagram which compares the forecast probabilities of having precipitation lower or equal than 5 mm with the observed frequencies of this event. For a perfect reliable forecast all points would be in the diagonal, so in this case ELR calibration clearly improves the performance of the raw forecast.

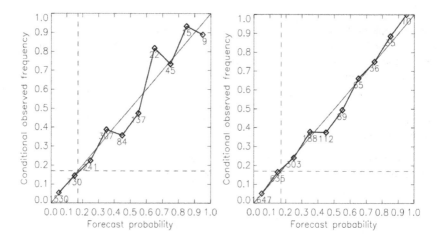

Figure 25. Reliability diagrams of H+126 area mean precipitation forecasts for the raw ensemble (left) and the ELR calibrated ensemble (right). In this case threshold $q = 5$ mm. From: Schmeits & Kok (2010). © American Meteorological Society. Reprinted with permission.

6.3. Non-homogeneous gaussian regression

The non-homogeneous Gaussian Regression (NGR) technique was proposed by Gneiting et al. (2005b). In its general form, the predictive PDF estimated by NGR is assumed to be a perfect Gaussian with the mean value being a bias-corrected weighted average of the ensemble members forecasts and the variance a linear function of the ensemble variance. That is (Gneiting et al., 2005b):

$$PDF(y) = \frac{1}{\sqrt{2\pi(c+ds^2)}} e^{-\frac{\left[y-(a+b_1x_1+\cdots+b_kx_k)\right]^2}{2(c+ds^2)}} \tag{17}$$

Here y is the weather quantity to be predicted, s is the ensemble spread and $\{x_1, ..., x_k\}$ are the k ensemble predictions of parameter y. $\theta = (a, b_1, ..., b_k, c, d)$ are the unknowns of this expression that have to be estimated by regression using the training data, which consist in a series of former forecast-observation pairs. The term non-homogeneous refers to the fact that the variances of the regression errors are not the same for all the values of $\{x_1, ..., x_k\}$ (they depend on s) as it is assumed in linear regression.

This way of representing the predictive PDF allows a natural understanding of the regression coefficients. Coefficient a is a bias-correction of the ensemble weighted mean. The weights $\{b_1, ..., b_k\}$ can be negative but for an easier interpretation Gneiting et al. (2005b) recommended constraining them to be non-negative which is done during the training process. On one hand side, they represent the performance of the ensemble members over the training period, with respect to the other members. On the other hand, they also reflect the correlations between ensemble members. Gneiting et al. (2005b) showed how a five members ensemble with three members using the same global data as initial and boundary conditions (so highly correlated) is automatically reduced to a three members ensemble after NGR calibration leaving only non-zero weights for the members that use different data as initial conditions. Variance coefficients c and d are constrained to be non-negative and they are a measure of the spread-skill relationship. For large values of d NGR variance is correlated to ensemble variance (s^2) so a significant correlation of the spread with the skill of the ensemble weighted mean is obtained. If spread and skill are independent of each other, d values will be negligible and it is c what represents the variance of the NGR calibrated mean.

Compared to other techniques (e.g. BMA; Raftery, 2005), NGR has the advantage that can be applied to ensembles whose members are non-distinguishable, such as the ECMWF ensemble prediction system (Molteni et al., 1996). In this case, NGR is simplified by constraining the $b_1 = ... = b_k$ coefficients to be equal, which at the same time agrees with the assumption of equiprobability of members. Now the analytical PDF (17) is reduced to:

$$PDF(y) = \frac{1}{\sqrt{2\pi(c+ds^2)}} e^{-\frac{\left[y-(a+bx_m)\right]^2}{2(c+ds^2)}} \tag{18}$$

Where x_m is the ensemble mean.

Due to the Gaussian shape of the analytical expression (Eq. 17), this technique is expected to be especially useful for weather parameters that have Gaussian distributions such as temperature or pressure. Figure 26 represents a real experiment by Hagedorn et al. (2008) where GFS (Toth & Kalnay, 1997), ECMWF and a multi-model (combining GFS and ECMWF) ensemble predictions of surface temperature are calibrated using NGR all else being equal. Continuous Ranked Probability Skill Score (CRPSS) is used as a performance measure for probabilistic forecasts (Jolliffe & Stephenson, 2003). The higher its values (closer to 1), the better the probabilistic forecast will be. When CRPSS reaches 0 it means that the probabilistic forecast has the same skill than the climatology (Hagedorn et al., 2008; see for details). In this case, it is clear the benefit of calibrating the ensembles with NGR.

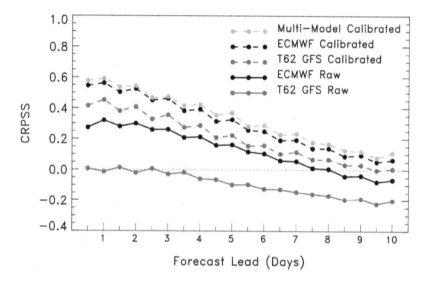

Figure 26. CRPSS of surface temperature forecasts with and without calibration. From: Hargedorn et al. (2008). © American Meteorological Society. Reprinted with permission.

A classical and widely extended technique for estimating the θ unknowns is maximum likelihood (Wilks, 2006). Nevertheless Gneiting et al. (2005b) demonstrated, for NGR probabilistic forecasts of temperature and surface pressure, that estimating θ by minimization of the continuous ranked probability score (Hersbach, 2000) gives clearly a better calibration of the PDF.

The adequate length of the training period for an operational approach is not unique. Raftery et al. (2005) showed that by using the previous 25 days the prediction intervals are the narrowest maintaining the right coverage of the verification (see the article for details). However, Wilks and Hamill (2007) used 45 days. An experimental study of the optimal length for one or more specific locations is desirable.

6.4. Ensemble dressing

The dressing technique is a statistical post-processing technique based on combining each member of a dynamical ensemble with its own statistical error ensemble.

Roulston & Smith (2003) proposed the use of a simple resampling scheme called *best member* method: individual members of an ensemble are *dressed* with an error distribution derived from the error made by the *best* member of the ensemble. The best member is defined as the member that is closer to the verification and the uncertainty of it is the one that is added to the rest of members. Identification of the best member is performed by means of multivariate forecasts although only univariate forecasts are dressed. The number of forecast variables required is estimated by looking at the fraction of the false best members (FBM). These FBM are defined using a distance on the vector space of the verification. If the N ensemble members are described as d-dimensional vectors, x_i *(i=1,...,N)* and y is the verification, the normalized distance is defined as (Roulston and Smith, 2003):

$$R_{i,d}^2 = \sum_{k=1}^{d} \frac{\left(x_{i,k} - y_k\right)}{\sigma_k^2}$$

(19)

Where d is the number of forecast variables being considered and $_k$ is the standard deviation of the kth component of the forecast vector. The best member is the one which the minimum $R_{i,d}^2$ although new additional variables are included. FBMs are the ones whose minimum distances are not maintained when new variables are added. The fraction of FBMs computed using the previous historic forecasts allows for a minimum number of variables required to obtain the best member to dress the dynamical ensemble. The best member error is also determined from historical ensemble forecasts by computing the differences between the best member and the corresponding verification.

Wang and Bishop (2005) showed by stochastic simulations that the best-member method can lead both to underdispersive or overdispersive ensembles. In addition to this, Wilks (2006) demonstrated that the dressed ensemble cannot be reliable. In order to alleviate these problems a new multivariate dressing method based on the second moment constraint is proposed. Ensemble bias is removed before building training statistics for the dressing kernel assuming that each ensemble is drawn for stochastic process. To dress the ensemble, statistical perturbations ε are added to each ensemble member. The covariance matrix Q is defined as (Wang & Bishop, 2005):

$$Q = <\varepsilon\varepsilon^T> = E\Omega E^T$$

(20)

Where the columns of E contain the eigenvectors of Q and the diagonal matrix contains the corresponding eigenvalues. Positive eigenvalues indicate that the ensemble is underdispersive in the directions of the corresponding eigenvectors and thus dressing is necessary. The Q matrix can be expressed as a function of the ensemble member forecasts and the verification values. The new dressing perturbation generator is defined as (Wang & Bishop, 2005):

$$\varepsilon = c_1 e_1^+ + c_2 e_2^+ + \cdots + c_l e_l^+$$

(21)

Where e_i^+, $i = 1, 2,..., I$, are the eigenvectors corresponding to the positive eigenvalues. The coefficients c_i are univariate random variables generated from a normal distribution with mean equal to zero and variance equal to the ith positive eigenvalue of Q.

The comparison of the original best-member dressing method with the second moment constraint dressing method confirms that the spread of the best-member dressed ensemble is indeed underdispersive, or even becomes overdispersive, depending on factors such as the undressed ensemble size, how underdispersive the undressed ensemble is and the nature of the subspace from which the best member is identified. On the other hand, for underdispersive ensembles, the second moment constraint dressing kernel correction always returns about the right amount of dispersion.

Although underdispersion is a common characteristic of an EPS, some variables have an overdispersive behavior (Feddersen & Andersen, 2005). Fortín et al. (2006) proposed to dress and weight each member differently to improve the reliability of the forecast and to correct the variable under or overdispersed. This method is very similar to BMA (Raftery et al., 2005) and today has only been applied to one-dimensional variables.

7. A brief description of some current state-of-the-art ensemble prediction systems

The Australian Bureau of Meteorology (BoM), Brazilian Centro de Previsao do Tempo e Estudos Climatico (CPTEC), China Meteorological Administration (CMA), the European Centre for Medium-Range Weather Forecasting (ECMWF), Japan Meteorological Agency (JMA), Korea Meteorological Administration (KMA), Meteorological Service of Canada (MSC), Météo-France (MF), UK Met Office (UKMO) and US National Centres for Environmental Prediction (NCEP), among others, run ensemble prediction systems.

The ECMWF-EPS is a global ensemble that is optimized for the medium range. It uses the singular vector technique (Ensemble Data Assimilation is under research and is starting to be used operationally together with SV) for providing the set of initial perturbations, as well as stochastic parameterizations to account for model errors. The ECMWF-EPS comprises 51 members with 62 vertical levels and a spectral horizontal resolution of T639.

In the NCEP ensemble the initial perturbations are obtained by the Ensemble Transform with Rescaling (ETR) technique. It also uses stochastic perturbations to account for model errors. It runs 20 members with 28 levels and a spectral horizontal resolution of T126.

The MetOffice ensemble, called MOGREPS, works with 24 members. It uses the ETKF (Ensemble Transform Kalman Filter) technique with scaling of perturbations using radiosonde and ATOVS observations (Bowler et al., 2008). The horizontal resolution is 0.83 degrees in longitude and 0.56 in latitude, and the number of vertical levels is 70.

The Japan Meteorological Agency (JMA) ensemble has 51 members. It uses singular vectors for the calculation of the initial perturbations and stochastic representation of physical parameterizations for accounting model error. The number of vertical levels is 60 levels and the spectral horizontal resolution is T319.

A source of information for studying global ensembles is the TIGGE project, which is a key component of the THORPEX Interactive Grand Global Ensemble, a World Weather Research

Programme for improving the accuracy of high-impact weather forecasts. In the TIGGE project, the forecasts of 10 global ensembles are archived, permitting the comparison of methods and results.

Limited area ensembles are developed for higher resolutions (nowadays from 2 km up to 25 km) and shorter time ranges (from 18 to 72 hours) than those of global ensembles. When a model can explicitly resolve convection (due to its characteristics and high resolution config-uration) it can represent more realistically typical precipitation patterns in the forecast field. However, convection forecasts (as well as other small scale processes) are very limited by their deterministic predictability (which is small due to its chaotic behaviour). Therefore, even in the short forecast range of only 24 hours, the prediction of details in the convection (such as location and timing of a thunderstorm) are usually very uncertain. Limited area ensembles can add information to deterministic high resolution forecasts and for this reason many operational weather centres are developing limited area ensembles.

Limited area ensembles running in operational centres are based on high resolution non-hydrostatic models, such as ALADIN (developed and maintained by a consortium of 16 National Meteorological Services, led by Météo-France), COSMO (developed by a consortium of seven NMS, led by Deutscher Wetterdienst), WRF (developed in the United States of America by NCAR, NOAA and others), HARMONIE (a model which shares code with ALADIN, developed by a consortium of 10 NMS) and the Unified Model (MetOffice). These ensembles can have an assimilation cycle which uses a wide class of meteorological observa-tions, in some cases including radar data trying to represent as much as possible actual precipitation processes. They need lateral boundary conditions typically coming from global ensembles or coarser limited area ensembles.

Just to mention two examples of limited area ensembles, we resume below the characteristics of the COSMO-DE ensemble (Deutscher Wetterdienst) and the AEMET-SREPS. Other opera-tional limited area ensembles are: the Norwegian targeted EPS LAMEPS (Frogner & Iversen, 2002), the Hungarian LAMEPS based on ALADIN (Hágel & Horányi, 2007), the multi-model GLAMEPS (ALADIN and HIRLAM consortium) and the limited area version of MOGREPS (MetOffice).

The COSMO-DE Ensemble is based on the convection resolving model COSMO-DE. It produces 2.8 km grid forecasts up to 18 hours, runs every 3 hours and assimilates estimated precipitation rates from RADAR.

The AEMET-SREPS uses the multi-model and multi-boundaries techniques for sampling initial, lateral conditions and model errors. It uses five global models for initial and lateral conditions and five limited area models running with every lateral and initial conditions (MM5, UM, HIRLAM, COSMO and HRM) thus producing 25 members. The horizontal resolution is 25 km and produces forecasts up to 72 hours twice every day. It covers a wide area (includes North Atlantic, Europe and North of Africa).

8. Conclusions and future directions

The most reliable and skilful theoretical forecasts from the current observed state of the atmosphere can be obtained through a Probability Distribution Function (PDF) which describes a comprehensive set of possible future states. The only practically feasible methodology to assess a forecasted PDF is using an Ensemble Prediction System (EPS), that is, a PDF sample of different but equally plausible Numerical Weather Prediction (NWP) forecasts (EPS members). Furthermore from the practical point of view EPS forecasts have been showed to be more reliable and skilful than a forecast from one single NWP model, even when the latter has a higher resolution.

This better performance is due to the fact that the set of non-linear equations which describe the future evolution of the atmosphere have a chaotic behaviour. This means that any uncertainty in the prediction process like two slightly different initial states could grow and lead to quite significantly distinct forecasted states. As a consequence, the predictability associated to any forecasted atmospheric state is always spatially and temporally limited but depending in each forecast on the uncertainty magnitude and the particular atmospheric situation.

The sources of errors and uncertainties which limit the predictability are mainly due to: a.) inaccuracies in the initial atmospheric state, estimated from available observations with their associated observational error and limited representativeness and imperfect assimilation systems, b.) inadequacies of the NWP models, related to dynamical NWP model formulation and physical parameterizations and c.) for Limited Area Model (LAM) EPSs, approximations and errors from Lateral Boundary Conditions. So any reliable and skilful EPS has to take into consideration all of these uncertainty sources by using different methodologies. Some of these methodologies are, for instance (and respectively): a.) singular vectors, bred vectors, Ensemble Transform Kalman Filter (ETKF) and Ensemble Data Assimilation (EDA), b.) multi-model, multi-physics or multi-parameterizations, multi-parameters and stochastic parameterizations, and c.) multi-boundaries.

In addition, as any forecasting system, EPS quality and value, that is the overall performance, has to be evaluated through objective verification, assessing the necessary and complementary set of properties (with the corresponding tools): consistency (rank histogram), reliability (spread-error and attributes diagrams), resolution (resolution component of Brier Score), discrimination (Relative Operating Characteristic curves), sharpness (Sharpness Histogram), skill (Brier Skill Score) and relative value (Relative Value Diagram). The main goal of verification, apart from assessing EPS forecast performance, is that EPS developers can be leaded to what and how to improve the EPS.

A number of EPS products, which take into account the forecast probabilities and the predictability concept, can serve the forecast guidance. The EPS products can be raw ones (e.g. stamps, plumes or spaghettis) or derived (e.g. ensemble mean and spread charts, probability and percentile maps, EPS-grams, clusters and extreme forecast indexes). Before the products production it would be advisable to calibrate the EPS in order to remove its systematic errors. Some statistical post-processing techniques for ensemble calibration are Bayesian Model Averaging, Logistic and Extended Logistic Regression, Non-homogeneous Gaussian Regression and Ensemble Dressing.

Regarding to future scenarios, the weather forecast process is expected to improve at all temporal scales, from the first hours to the climatic scales, and to be done at finer spatial scales, due to the fact of having more and better observations, better NWP models, increased supercomputer resources, etc. Even though it is expected a reduction of errors and uncertainties, they will be always present and limit the predictability. This means that the majority of ideas and methodologies explained in this chapter are going to be useful for the next generation of weather forecasting systems, although new ones are expected to be developed. Then some of the future directions of work are outlined.

From the point of view of the current experience in EPS development, the multi-model and multi-analysis (from independent Global NWP models) approach, has showed to have better performance than any theoretical methodology based on a single model. This fact means that there is not enough knowledge about the different model uncertainties and that there can even be other unknown sources of error. Anyway, the latter approach is expected to be used intensively in the next EPS generations and even to overcome the former approach.

On the other hand, the better the EPS performance seems to be, the more number of error sources are considered and even the more methodologies are used together. Thus, future EPS developments are going to follow this line, partly because it increases the spread counteracting the common EPS shortcoming, the underdispersion. Anyway particular attention has to be paid not to increase spuriously the EPS spread, that is, without increasing the skill.

As it has been mentioned before, other EPS developments will come from having better assimilation techniques, better generation of initial conditions and better methodologies to tackle model errors and uncertainties. This fact means that, in a foreseeable future, EPSs will become more complex.

Finally in the current and the next decade there will be an important increase in the horizontal and vertical resolutions of the EPSs linked to the NWP model developments for smaller grid spacing. Thus, spatial resolutions of the next Global and LAM EPSs generations are going to be, respectively, close to the non-hydrostatic scale (e.g. 8-16 km), and about the meso-gamma or convection-resolving scale (e.g. 1-4 km). One consequence of this will be that verification will have to evolve to an objected-oriented way (e.g. SAL or MODE techniques). Because of uncertainties grow faster as the resolved scales are smaller, due to a more intrinsic chaotic nature, another consequence will be that the only feasible methodology to forecast the weather at these scales will be ensemble forecasting.

Author details

Alfons Callado, Pau Escribà, José Antonio García-Moya, Jesús Montero, Carlos Santos, Daniel Santos-Muñoz and Juan Simarro

Agencia Estatal de Meteorología (AEMET), Spain

References

[1]   Anderson, J. L. (1996). A method for producing and evaluating probabilistic forecasts from ensemble model integrations. *Journal of Climate*, 9, pp. 1518–1530.

[2]   Andersson, E. & Coauthors. (1998). The ECMWF implementation of the three-dimensional variational assimilation (3D-Var). III: Experimental results. *Quarterly Journal of the Royal Meteorological Society*, 124, pp. 1831-1860.

[3]   Athens, R. & Warner, T. (1978). Development of hydrodynamic models suitable for air pollution and other mesometeorological studies. *Monthly Weather Review*, 106, pp. 1045-1078.

[4]   Bishop, C. H., Etherton, B. J. & Majumdar S. J. (2001). Adaptive sampling with the ensemble transform Kalman filter. Part I: Theoretical aspects. *Monthly Weather Review*, 129, pp. 420–436

[5]   Bowler, N. E. (2008). Accounting for the effect of observation errors on verification of MOGREPS. *Meteorological Applications.*, 15, pp. 199-205.

[6]   Bradley A. A., Schwartz S. S. & Hashino T. (2008). Sampling Uncertainty and Confidence Intervals for the Brier Score and Brier Skill Score. *Weather and Forecasting*, 23:5, pp. 992-1006.

[7]   Buizza, R. & Palmer, T. (1995). The singular vectors structure of the atmospheric general circulation. *Journal of Atmospheric Sciences*, 52, pp. 1434-1456.

[8]   Buizza, R. & Palmer, T. (1997). Potential forecast skill of ensemble prediction, and spread and skill distributions of the ECMWF Ensemble Prediction System. *Monthly Weather Review*, 125, pp. 99-119.

[9]   Buizza, R., Petroliagis, T., Palmer, T. N., Barkmeijer, J., Hamrud, M., Hollingsworth, A., Simmons, A. & Wedi, N. (1998). Impact of model resolution and ensemble size on the performance of an ensemble prediction system. *Quarterly Journal of the Royal Meteorological Society*, 124, pp. 1935-1960.

[10] Buizza, R., Miller, M. & Palmer, T. (1999). Stochastic representation of model uncertainties in the ECMWF Ensemble Prediction System. *Quarterly Journal of the Royal Meteorological Society*, 125, pp. 2887-2908.

[11] Buizza, R., Leutbecher, M., & Isaksen, L. (2008). Potential use of an ensemble of analyses in the ECMWF Ensemble Prediction System. *Quarterly Journal of the Royal Meteorological Society*, 134, pp. 2051-2066.

[12] Candille, G., & Talagrand, O. (2005). Evaluation of probabilistic prediction systems for a scalar variable. *Quarterly Journal of the Royal Meteorological Society*, 131, pp. 2131–2150.

[13] Candille, G. & Talagrand, O. (2008). Impact of observational error on the validation of ensemble prediction systems. *Quarterly Journal of the Royal Meteorological Society*, 134, pp. 959-971.

[14] Casati, B., Wilson, L. J., Stephenson, D. B., Nurmi, P., Ghelli, A., Pocernich, M., Damrath, U., Ebert, E. E., Brown, B. G. & Mason, S. (2008). Forecast verification: current status and future directions. *Meteorological Applications*, 15, pp. 3–18. doi: 10.1002/met. 52.

[15] Charney, J. G., Fjørtoft, R. & von Neumann, J. (1950). Numerical integration of the barotropic vorticity equation. *Tellus*, 2, pp. 237-254.

[16] Cherubini, T., Ghelli, A., & Lalaurette, F. (2002). Verification of precipitation forecasts over the Alpine region using a high-density observing network. *Weather and Forecasting*, 17, pp. 238–249.

[17] Clark, A. J., Gallus, W. A., Xue, M. & Kong, F. (2009). A comparison of precipitation forecast skill between small convection-allowing and large convection-parameterizing ensembles. *Weather and Forecasting*, 24, pp. 1121–1140.

[18] Daley, R. (1991). *Atmospheric Data Analysis*. Cambridge University Press, Cambridge.

[19] Davis, C. A., Brown, B. G., Bullock, R. & Halley-Gotway, J. (2009). The Method for Object-Based Diagnostic Evaluation (MODE) Applied to Numerical Forecasts from the 2005 NSSL/SPC Spring Program. *Weather and Forecasting*, 24, pp. 1252–1267, doi: 10.1175/2009WAF2222241.1

[20] Dempster, A. P., Laird, N. M., & Rubin, D. B. (1977). Maximum likelihood for incomplete data via the EM algorithm (with discussion). *Journal of the Royal Statistical Society*, Ser. B, 39, pp. 1–38.

[21] Du, J. & Tracton, M. (2001). Implementation of a real-time short range ensemble forecasting system at NCEP: An update. Preprints, *Ninth Conference on Mesoscale Processes*, pp. 355-360. Ft. Laurderdale, FL: American Meteorological Society.

[22] Ebert, E. E. (2001). Ability of a Poor Man's Ensemble to Predict the Probability and Distribution of Precipitation. *Monthly Weather Review*, 129, pp. 2461–2480.

[23] Ebert, E. E., Gallus, W. A. (2009). Toward Better Understanding of the Contiguous Rain Area (CRA) Method for Spatial Forecast Verification. *Weather and Forecasting*, 24, pp. 1401–1415.

[24] Ebisuzaki, W., & Kalnay, E. (1991). Ensemble experiments with a new lagged average forecasting scheme. *WMO Research Activities in Atmospheric and Oceanic Modeling* Rep. 15, 308 pp.

[25] Eckel, F. A. & Mass, C. F. (2005). Aspects of effective mesoscale, short-range ensemble forecasting, *Weather and Forecasting*, 20, pp. 328-350.

[26] Efron, B. & Tibshirani, R. (1997). Improvements on Cross-Validation: The 632+ Bootstrap Method. *Journal of the American Statistical Association*, 92, No. 438, pp. 548-560.

[27] Ehrendorfer, M. (1997). Predicting the uncertainty of numerical weather forecasts: a review. *Meteorologische Zeitschrift*, N.F. 6, pp. 147-183.

[28] Emanuel, K. A. (1979). Inertial Instability and Mesoscale Convective Systems. Part I: Linear Theory of Inertial Instability in Rotating Viscous Fluids. *Journal of the Atmospheric Sciences*, 36, pp. 2425-2449.

[29] Escribà, P., Callado, A., Santos, D., Santos, C., García-Moya, J.A. & Simarro, J. (2010). Probabilistic prediction of raw and BMA calibrated AEMET-SREPS: the 24 of January 2009 extreme wind event in Catalunya. *Advances in Geosciences*, 26, pp. 119-124.

[30] Evans, R. E., Harrison, M. & Graham, R. (2000). Joint medium range ensembles from the Met. Office and ECMWF systems. *Monthly Weather Review* , 128, pp. 3104-3127.

[31] Feddersen, H. & Andersen, U. (2005). A method for statistical downscaling of seasonal ensemble predictions. *Tellus*, 57A, pp. 398–408.

[32] Ferranti, L. & Corti, S. (2010). Ensemble prediction skill in relation with large scale circulation patterns. *EMS Annual Meeting Abstracts*, Vol. 7, EMS2010-769, 2010, 10th EMS / 8th ECAC

[33] Ferro, C. A. T. (2007a). A Probability Model for Verifying Deterministic Forecasts of Extreme Events. *Weather and Forecasting*, 22, pp. 1089–1100.

[34] Ferro., C. A. T. (2007b). Comparing Probabilistic Forecasting Systems with the Brier Score. *Weather and Forecasting* 22:5, pp. 1076-1088.

[35] Ferro, C. A. T., Richardson, D. S. & Weigel, A. P. (2008). On the effect of ensemble size on the discrete and continuous ranked probability scores. *Meteorological Applications*, 15, pp. 19–24.

[36] Fortín, V., Favre, A. C. & Saïd, M. (2006). Probabilistic forecasting from ensemble prediction systems: Improving upon the best-member method by using a different weight and dressing kernel for each member. *Quarterly Journal of the Royal Meteorological Society*, 132, pp. 1349–1369.

[37] Frogner, I. L. & Iversen, T. (2002). High-resolution limited-area ensemble predictions based on low-resolution targeted singular vectors. *Quarterly Journal of the Royal Meteorological Society*, 128, pp. 1321–1341.

[38] García-Moya, J. A., Callado, A., Escribà, P., Santos, C., Santos-Muñoz, D. & Simarro, J. (2011). Predictability of short-range forecasting: a multimodel approach. *Tellus A*, 63, pp. 550–563.

[39] Ghelli, A. & Lalaurette, F. (2000) Verifying precipitation forecasts using upscaled observations. *ECMWF Newsletter* 87, ECMWF, Reading, United Kingdom, pp. 9–17.

[40] Ghelli, A. & Primo, C. (2009). On the use of the extreme dependency score to investigate the performance of a NWP model for rare events. *Meteorological Applications.*, 16, pp. 537–544.

[41] Gilleland, E., Ahijevych, D., Brown, B. G., Casati, B. & Ebert, E. E. (2009). Intercomparison of Spatial Forecast Verification Methods. *Weather and Forecasting*, 24, pp. 1416–1430.

[42] Gneiting, T. & Raftery, A. E. (2005a). Weather forecasting with ensemble methods. *Science* , 310, pp. 248-249.

[43] Gneiting, T., Raftery A.E., Westveld III, A.H. & Glodman, T. (2005b). Calibrated Probabilistic Forecasting Using Ensemble Model Output Statistics and Minimum CRPS Estimation. *Monthly Weather Review*. 133, pp. 1098-1118.

[44] Grimit, E. & Mass, C. (2002). Initial results of a mesoscale short-range ensemble forecasting system over the Pacific Northwest. *Weather and Forecasting* , 17, pp. 192-205.

[45] Gutiérrez, J. M., Cofiño, A. S., Cano, R. & Rodríguez, M. A. (2004). Clustering methods for statistical downscaling in short-range weather forecasts. *Monthly Weather Review*, 132, pp. 2169–2183.

[46] Hagedorn, R., Hamill, T. M., & Whitaker, J. S. (2008). Probabilistic Forecast Calibration using ECMWF and GFS Ensemble Reforecasts. Part I: Two-meter temperatures. *Monthly Weather Review*, 136, pp. 2608-2619.

[47] Hágel & Horányi (2007). The ARPEGE/ALADIN limited area ensemble prediction system: the impact of global targeted singular vectors. *Meteorologische Zeitschrift*, 16, Number 6, December 2007, pp. 653-663.

[48] Hamill, T. & Colucci, S. (1997). Verification of ETA-RSM short-range ensemble forecast. *Monthly Weather Review*, 125, pp. 1312-1327.

[49] Hamill, T. & Colucci, S. (1998). Evaluation of ETA-RSM probabilistic precipitation forecasts. *Monthly Weather Review*, 126, pp. 711-724.

[50] Hamill, T., Snyder, C. & Morss, R. (2000). A comparison of probabilistic forecasts from bred, singular-vector, and perturbed observation ensembles. *Monthly Weather Review*, 128, pp. 1835-1851.

[51] Hamill, T. M., Hagedorn R. & Whitaker, J. S. (2008). Probabilistic Forecast Calibration using ECMWF and GFS Ensemble Reforecasts. Part II: Precipitation. *Monthly Weather Review*. 136, pp. 2620-2632.

[52] Hersbach, H. (2000). Decomposition of the continuous ranked probability score for ensemble prediction systems. *Weather and Forecasting,* 15, pp. 559-570.

[53] Hoffman, R. N. & Kalnay, E. (1983). Lagged average forecasting, an alternative to Monte Carlo forecasting. *Tellus A,* 35, pp. 100–118.

[54] Hollingsworth, A. (1980). An experiment in Montecarlo Forecasting. *Workshop on Stochastic-Dynamic Forecasting,* pp. 65-85. Reading, United Kingdom.

[55] Hohenegger, C. & Schär, C. (2007). Predictability and error growth dynamics in cloud-resolving models. *Journal of Atmospheric Sciences,* 64, pp. 4467-4478.

[56] Hoeting, J. A., Madigan, D. M., Raftery, A. E. & Volinsky, C. T. (1999). Bayesian model averaging: A tutorial (with discussion). *Statistical Science,* 14, pp. 382–401.

[57] Hou, D., Kalnay, E. & Droegemeier, K. (2001). Objetive verification of the SAMEX'98 ensemble forecast. *Monthly Weather Review,* 129, pp. 73-91.

[58] Houtekamer, P. L., Lefaivre, L., Derome, J., Ritchie, H. & Michell, H. (1996). A system simulation approach to ensemble prediction. *Monthly Weather Review,* 124, pp. 1225-1242.

[59] Houtekamer, P. L. & Mitchell, H. L. (1998). Data assimilation using an ensemble Kalman filter technique. *Monthly Weather Review,* 126, pp. 796-811.

[60] Jolliffe, I. T. & Stephenson, D. B. (2003). *Forecast Verification: A Practitioner's Guide in Atmospheric Science.* Wiley, New York.

[61] Jones, M., Colle, B. & Tongue, J. (2007). Evaluation of a mesoscale short-range ensemble forecast system over the Northeast United States. *Weather and Forecasting* , 22, pp. 36-55.

[62] Kalnay, E. & Ham, M. (1989). Forecasting forecast skill in the Southern Hemisphere. Extended Abstracts. *Third Int. Conf. on Southern Hemisphere Meteorology and Oceanography,* pp. 24-27. Buenos Aires, Argentina.: American Meteorological Society.

[63] Kass, R. E., & Raftery, A. E. (1995). Bayes factors. *Journal of the American Statistical Association,* 90, pp. 773–795.

[64] Kharin, V. V. & Zwiers, F. W. (2003). On the ROC score of probability forecasts. *J. Clim.* 16, pp. 4145–4150.

[65] Krishnamurthy, V. (1993). A Predictability Study of Lorenz's 28-Variable Model as a Dynamical System. *Journal of the Atmospheric Sciences,* Vol. 50, No. 14, pp.2215-2229.

[66] Krishnamurti, T. N., Kishtawal, C., LaRow, T., Bachiochi, D., Zhang, Z., Willford, C., et al. (1999). Improved weather and seasonal climate forecast from multimodel superensemble. *Science* , 285, pp. 1548-1550.

[67] Lalaurette, F. (2003) Early detection of abnormal weather conditions using a probabilistic extreme forecast index. *Quarterly Journal of the Royal Meteorological Society*, 129, pp. 3037-3057.

[68] Leamer, E. E. (1978). *Specification Searches*. Wiley, New York.

[69] Leith, C. E. (1974). Theoretical skill of Monte Carlo forecast. *Monthly Weather Review*, 102, pp. 409-418.

[70] Lin, J. W. B. & Neelin, J. D. (2002). Considerations for stochastic convective parameteriza-tion. *Journal of Atmospheric Sciences*, Vol. 59, No. 5, pp. 959-975.

[71] Lorenz, E. N. (1963). Deterministic nonperiodic flow. *Journal of Atmospheric Sciences.*, 20, pp. 130-141.

[72] Lorenz, E. N. (1969). The predictability of a flow which possesses many scales of motion. *Tellus*, 21, pp. 1-19.

[73] Lyapunov, A. M. (1992). *The general problem of the stability of motion*, Translated by A. T. Fuller, London: Taylor & Francis, ISBN 978-0748400621

[74] Lynch, P. (2006). "The ENIAC Integrations". The Emergence of Numerical Weather Prediction. *Cambridge University Press*, pp. 206-208, ISBN 9780521857291.

[75] Mason, S. J. (2004). On using "climatology" as a reference strategy in the Brier and ranked probability skill scores. *Monthly Weather Review*, 132, pp. 1891-1895.

[76] McLachlan, G. J. & Krishnan, T. (1997). *The EM Algorithm and Extensions*. Wiley, 274 pp.

[77] Mechoso, C. R. & Arakawa, A. (2003). *General ciculation. Models*. In J. R. Holton, J. A. Curry and J. A. Pyle, Encyclopedia of Atmospheric Sciences, pp. 861-869. Oxford: Academic Press.

[78] Mesinger, F., Janjic, Z., Nickovic, S., Gavrilov, D. & Deaven, D. G. (1988). The step mountain coordinate: model description and performance for cases of alpine cyclogenesis and for a case of an Appalachian redevelopment. *Monthly Weather Review*, 116, pp. 1493-1518.

[79] Molteni, F., Buizza, R., Palmer, T. & Petroliagis, T. (1996). The ECMWF Ensemble Prediction System: Methodology and Validation. *Quarterly Journal of the Royal Meteorological Society*, 122, pp. 73-119.

[80] Mullen, S. & Baumhefner, P. (1989). The impact of initial condition uncertainty on numerical simulations of large-scale explosive cyclogenesis. *Monthly Weather Review*, 117, pp. 2800-2821.

[81] Murphy, J. M. (1988). The impact of ensemble forecasts on predictability. *Quarterly Journal of the Royal Meteorological Society*, 114, pp. 463-493.

[82] Murphy, A. H. (1993). What is a good forecast?, An essay on the nature of goodness in weather forecasting. *Weather and Forecasting*, 8, pp. 281–293.

[83] Murphy, J., Sexton, D., Barnett, D., Jones, D., Webb, M., Collins, M. & Stainforth, D. (2004). Quantification of modelling uncertainties in a large ensemble of climate change simulations. *Nature*, 430, pp. 768–772.

[84] PaiMazumder, D. & Mölders, N. (2009). Theoretical assessment of uncertainty in regional averages due to network density and design. *Journal of Applied Meteorology and. Climatology*, 48, pp. 1643-1666.

[85] Palmer, T. N., Mureau, R., Buizza, R., Chapelet, P. & Tribbia, J. (1992). Ensemble prediction. *ECMWF Research Department Technical Memorandum*, 188, 45 pp.

[86] Palmer, T. N. (1997). On parameterizing scales that are only somewhat smaller than the smallest resollved scales, with application to convection and orography. *Proceedings of the ECMWF Workshop on New Insights and Approaches to Convective Parameterization*, 4-7 November 1996, pp. 328-337.

[87] Palmer, T. N. (2001). A nonlinear dynamical perspective on model error: Aproposal for non-local stochastic-dynamic parametrization in weather and climate prediction models. *Quarterly Journal of the Royal Meteorological Society*, 127, pp. 279–304.

[88] Palmer, T., Alessandri, A., U.Andersen, Cantelaube, P., Davey, M. et al. (2004). Development of a European multimodel ensemble system for seasonal to inter-annual prediction (DEMETER). *Bulletin of the American Meteorological Society*, 85, pp. 853-872.

[89] Pellerin, G., Lefaivre, F., Houtekamer, P. & Girard, C. (2003). Increasing the horizontal resolution of ensemble forecast at CMC. *Nonlinear Processes in Geophysics*, 10, pp. 463-468.

[90] Persson, A., Grazzini, F. (2005). User guide to ECMWF forecast products. Meteorological Bulletin M3.2, ECMWF, Reading, United Kingdom, 115 pp.

[91] Raftery, A. E., Balabdaoui, F., Gneiting, T. & Polakowski, M. (2005). Using Bayesian Model Averaging to calibrate forecast ensembles. *Monthly Weather Review*, 133, pp. 1155-1174.

[92] Richardson, D. S. (2000). Skill and relative economic value of the ECMWF ensemble prediction system. *Quarterly Journal of the Royal Meteorological Society*, 126, pp. 649–667.

[93] Roebber, P. J. & Reuter, G. W. (2002). The sensitivity of precipitation to circulation details. Part II: Mesoscale modeling. *Monthly Weather Review*, 130, pp. 3–23.

[94] Rodwell, M. J., Richardson, D. S., Hewson, T. D. & Haiden, T. (2010). A new equitable score suitable for verifying precipitation in numerical weather prediction. *Quarterly Journal of the Royal Meteorological Society*, 136, pp. 1344–1363.

[95] Roulston, M. S. & Smith, L. A. (2003). Combining dynamical and statistical ensembles. *Tellus* A, 55, pp. 16–30.

[96] Saetra, Ø., Hersbach, H., Bidlot, J. R. & Richardson, D. S. (2004). Effects of Observation Errors on the Statistics for Ensemble Spread and Reliability. *Monthly Weather Review*, 132, pp. 1487-1501.

[97] Santos, C. & Ghelli, A. (2011). Observational probability method to assess ensemble precipitation forecasts. *Quarterly Journal of the Royal Meteorological Society*, doi: 10.1002/qj.895.

[98] Schmeits, M. J & Kok, K. J. (2010). A comparison between Raw Ensemble Output, (Modified) Bayesian Model Averaging, and Extended Logistic Regression using ECMWF Ensemble Precipitation Reforecasts. *Monthly Weather Review,* 138, pp. 4199-4211.

[99] Shutts, G., & Palmer, T. N. (2004). The use of high resolution numerical simulations of tropical circulation to calibrate stochastic physics scheme. *Proceeding ECMWF/ CLIVAR Workshop on Simulation and Prediction of Intra-Seasonal Variability with Emphasis on the MJO*, Reading, United Kingdom, ECMWF, pp. 83-102.

[100] Shutts, G. (2005). A kinetic energy backscatter algorithm for use in ensemble prediction systems. *Quarterly Journal of the Royal Meteorological Society*, 131, pp. 3079–3102.

[101] Sloughter, J. M., Raftery, A. E., Gneiting T. & Fraley C. (2007). Probabilistic quantitative precipitation forecasting using Bayesian model averaging. *Monthly Weather Review*, 135, pp. 3209–3220.

[102] Sloughter, J.M., Gneiting, T. & Raftery, A.E. (2010). Probabilistic Wind Speed Forecasting using Ensembles and Bayesian Model Averaging. *Journal of the American Statistical Association*, 105, pp. 25-35.

[103] Stensrud, D., Bao, J. W. & Warner, T. (1998). Ensemble forecasting of mesoscale convective systems. In A. M. Soc (Ed.), *12th Conference on Numerical Weather Prediction, Phoenix , AZ*. Preprints, pp. 265-268.

[104] Stensrud, D., Brooks, H., Du, J., Tracton, M. & Rogers, E. (1999). Using ensembles for short-range forecasting. *Monthly Weather Review*, 127, pp. 433-446.

[105] Stensrud, D. J. & Yussouf, N. (2007). Reliable probabilistic quantitative precipitation forecasts from a short-range ensemble forecast system. *Weather and Forecasting*, 22, pp. 3–17.

[106] Stephenson, D. B., Casati, B., Ferro, C. A. T. & Wilson, C. A. (2008). The extreme dependency score: a non-vanishing measure for forecasts of rare events. *Meteorological Applications*, 15, pp. 41–50.

[107] Toth, Z. & Kalnay, E. (1993). Ensemble forecasting at NMC: The generation of perturbations. *Bulletin of the American Meteorological Society*, 74, pp. 2317-2330.

[108] Toth, Z. & Kalnay, E. (1997). Ensemble forecasting at NCEP: The breeding method. *Monthly Weather Review*, 125, pp. 3297-3318.

[109] Tracton, M. S. & Kalnay, E. (1993). Operational ensemble forecasting prediction at the National Meteorological Centre: Practical aspects. *Weather and Forecasting*, 8, pp. 379-398.

[110] Tribbia, J. J. & Baumhefner, D. P. (1988). The reliability of improvements in deterministic short-range forecasts in the presence of initial state and modeling deficiencies. *Monthly Weather Review*, 116, pp. 2276-2288.

[111] Wandishin, M., Mullen, S., Stensrud, D. & Brook, H. (2001). Evaluation of a short-range multimodel ensemble system. *Monthly Weather Review*, 129, pp. 729-747.

[112] Wang, X. & Bishop, C. H. (2003). A comparison of breeding and ensemble transform Kalman filter ensemble forecast schemes. *Journal of Atmospheric Sciences*, 60, pp. 1140–1158.

[113] Wang, X. & Bishop, C. H. (2005). Improvement of ensemble reliability with a new dressing kernel. *Quarterly Journal of the Royal Meteorological Society*, 131, pp. 965–986

[114] Wernli, H., Paulat, M., Hagen, M., Frei, C. (2008). SAL — A Novel Quality Measure for the Verification of Quantitative Precipitation Forecasts. *Monthly Weather Review*, 136, pp. 4470–4487.

[115] Whitaker, J. S. & Loughe, A. F. (1998). The relationship between ensemble spread and ensemble mean skill. *Monthly Weather Review*, 126, pp. 3292–3302.

[116] Wilks, D. S., (2006). *Statistical Methods in the Atmospheric Sciences*. 2nd ed. Academic Press, 648 pp.

[117] Wilks, D. S. & Hamill, T.M. (2007). Comparison of ensemble-MOS methods using GFS reforecasts. *Monthly Weather Review*. 135, pp. 2379-2390.

[118] Wilks, D. S. (2009). Extending logistic regression to provide full-probability-distribution MOS forecasts. *Meteorological Applications*, 16, pp. 361-368.

[119] Williamson, D. L. (2007). The evolution of dynamical cores for global atmospheric models. *Journal of the Meteorological Society of Japan*, 85B, pp. 241-269.

[120] Wilson, L. J., Beauregard, S., Raftery, A. E. & Verret, R. (2007). Calibrated Surface Temperature Forecasts from the Canadian Ensemble Prediction y stem Using Bayesian Model Averaging (with discussion). *Monthly Weather Review*, 135, pp. 1364-1385. Discussion pages 4226-4236.

[121] Wobus, R. & Kalnay, E. (1995). Three years of operational prediction of forecast skill. *Monthly Weather Review*, 123, pp. 2132-2148.

[122] Ziehmann, C. (2000). Comparison of a single-model EPS with a multimodel ensemble consisting of a few operational models. *Tellus*, 52A, pp. 280–299.

[123] Zhang, F. (2005). Dynamics and structure of mesoscale error covariance of a winter cyclone estimated through short-range ensemble forecasts. *Monthly Weather Review*, 133, pp. 2876-2893.

Chapter 2

Measurements and Observations of Meteorological Visibility at ITS Stations

Nicolas Hautière, Raouf Babari, Eric Dumont,
Jacques Parent Du Chatelet and Nicolas Paparoditis

Additional information is available at the end of the chapter

1. Introduction

In the presence of dust, smoke, fog, haze or pollution, meteorological visibility is reduced. This reduction constitutes a common and vexing transportation problem for different public authorities in multiple countries throughout the world.

First, low visibility is obviously a problem of traffic safety. Road crashes which occur in fog are generally more severe as the average crash [1]. According to NOAA [2], in the United States there are approximately 700 annual fog-related fatalities, defined as occurring when visibility is less than a $\frac{1}{4}$ mile. Fog constitutes an equally important issue in France, a smaller country, with over 100 annual fatalities attributed to low visibility, defined as occurring when visibility is less than a 400 meters ($\approx \frac{1}{4}$ mile). Indeed, fog causes similar and significant problems on Northern America and French highways. The combination of fog and smoke presence on a motorway was the cause of dramatic pile-ups in France, e.g. on the A10 in 2002 near Coulombiers (58 vehicles involved, 40 injured, 8 deaths). Indeed, even if the origin of both phenomena differs, the combination of their mutual effect on the visibility is exponential, which leads to close to zero visibility areas. It should be stressed that the solution lies not necessarily in better low visibility detection but in drivers' response to fog that they encounter. Indeed, the behavior of drivers in fog is often inappropriate (e.g., reduced headways, altered reaction times) but to understand the origins of these dangerous behaviors is difficult [3]. Different countermeasures have been tested to mitigate the impact of critically reduced visibility [4]. The California San Joaquin and Sacramento Valley regions are particularly adequate test-beds for such measures, because of the well-known Tule fog phenomenon. In the Stockton area of Caltrans District 10, the Caltrans Automated Warning System (CAWS) employs roadside weather stations and visibility meters to provide automated detection [5]. In District 6, Caltrans has installed the "Fog Pilot" system, which provides a high-technology solution every $\frac{1}{4}$ mile along a 12-mile (20-km) portion of State Route 99.

In addition to the safety problem, reduced visibility is the cause of delays and disruption in air, sea and ground transportation for passengers and freight. On freeways, massive pile-ups create exceptional traffic congestions which sometimes force the operator to momentarily close the road. Fog-related road closures are not an uncommon subject for news headlines. For example, the Heathrow airport was blocked for three days during the 2006 Christmas time. Such events have of course important economic impacts [6]. According to [7], in 1974 fog was estimated to have cost over roughly £120 millions at 2010 prices on the roads of Great Britain. This number includes the cost of medical treatment, damage to vehicles and property, as well as the administrative costs of police, services and insurance, but they do not include the cost of delays to vehicle passengers not directly involved in the accident.

An ability to accurately monitor visibility helps resolve these problems. Important transportation facilities where safety is critical, such as airports, are generally instrumented for monitoring visibility with devices that are expensive and hence, scarce. Cost is precisely the reason why highway meteorological stations are seldom equipped with visibility metering devices. In this context, using already existing and ubiquitous highway cameras is of great interest, as these are low cost sensors already deployed for other purposes such as traffic monitoring [8]. Furthermore, introducing new functionalities into roadside cameras will make them multipurpose and thus more cost-effective, easing their deployment along the roads. In the United States, this potential has been identified by US DOT and was evaluated in the CLARUS Initiative [9], and these efforts may continue with the US DOT IntelliDrive program. In France, a similar initiative has been launched between Ifsttar (French institute of science and technology for transport, development and networks), Météo France (French National Weather Service) and IGN (French National Geographical Institute), three public research institutes dealing with road operation, weather monitoring and forecasting, and geography and cartography, respectively. The French initiative aims at assessing the potential of highway cameras to monitor visibility for different applications ranging from safety hazard detection to air quality monitoring. In the future, such initiatives might make it possible to monitor visibility reduction at the scale of a road itinerary. Prediction, which will soon be possible for airports [10], might even be envisioned.

2. Objectives

2.1. Problematic

Reduced visibility in the atmosphere is directly related to light scattering by air molecules and airborne particles. This tenet of physics is the basis of the operating principle of visibility meters. There are two types of instruments for measuring atmospheric visibility: transmissometers and scatter meters. The transmissometer extrapolates the attenuation of a light beam emitted from a source to a receiver at a known path length in order to estimate the distance for which the emitted light is attenuated by 95%. The transmissometer is also used to calibrate the scatter meter. A scatter meter assesses the dispersion of a light beam at a particular scattering angle, more often close to 40°(forward-scatter meters). Visibility meters can measure the meteorological visibility distance up to a few tens of kilometers with an error of 20%. The annual statistics on fog occurrence in France, i.e. episodes where the meteorological visibility distance is lower than 1000 meters are obtained from 60 weather stations distributed over the entire territory. Today, about 160 meteorological stations with visibility measurements are available in real time. But as such, this data cannot be used for predicting fog events and warning road authorities and hence drivers. Indeed, the local nature of this phenomenon is not compatible with the current capacity of meteorological agencies to monitor it accurately. Some studies seek to exploit the photosensitive cells of fixed cameras to measure the visibility.

2.2. The potential of CCTV networks

A survey has been conducted by Météo France on the French motorway networks to estimate the potential of existing CCTV networks to observe the visibility: In 2009, the French motorway network was 8,372 km long and was equipped with approximately 2,000 cameras. Accounting the fact that some are grouped together and some are dedicated to tunnel safety, a potential of 1,000 cameras available to monitor the weather was estimated. The French highway network is also equipped with cameras but they are less numerous. This whole network covers the territory quite uniformly. Consequently, a roadside sensor network constitutes a relevant mesh able to feed meteorological centers with geolocalized data.

2.3. Intelligent Transportation System

The term Intelligent Transportation Systems (ITS) refers to information and communication technology (applied to transportation infrastructure and vehicles) that improves transportation outcomes such as transportation safety, transportation productivity, travel reliability, informed travel choices, social equity, environmental performance and network operation resilience. The recent development of real-time data exchange systems between vehicles and infrastructure allows linking operation centers, roadside sensors and vehicles by means of so-called "ITS Stations". Such technology fosters a new generation of Intelligent Transportation Systems. The objectives of this work is thus to design computer vision methods, which can be implemented into camera-based surveillance systems connected to ITS Stations, in order to detect and characterize reduced visibility conditions, so as to mitigate the risk of accidents by alerting the drivers or by computing adaptive speeds related to the offered visibility distance.

3. Background on meteorological visibility

3.1. Visibility sensor requirements

According to [11, 12], the "meteorological visibility distance" denotes the greatest distance at which a black object of a suitable dimension can be seen in the sky on the horizon, with a threshold contrast set at 5%. The meteorological visibility distance is thus a standard dimension which characterizes the opacity of the atmosphere. According to [13], the road visibility is defined as the horizontal visibility determined 1.2 m above the roadway. It may be reduced to less than 400 m by fog, precipitations or projections. Four visibility ranges are defined and are listed in Tab. 1. Based on these definitions, a visibility sensor should assign the visibility range to one of the four categories and detect the origin of the visibility reduction, i.e. it should detect fog, rain and projections. In this section, the focus is on daytime fog detection and visibility range estimation using two complementary systems.

Visibility range index	Horizontal visibility distance (m)
1	200 to 400
2	100 to 200
3	50 to 100
4	< 50

Table 1. Ranges issued from the French standard NF P 99-320 on highway meteorology, in agreement with the international practice.

3.2. Vision through the atmosphere

The apparent luminance of the road pavement L is given by Koschmieder's law [14] which adds to Beer-Lambert's law a second term corresponding to the atmospheric veil:

$$L = L_0 e^{-kd} + L_f(1 - e^{-kd}) \tag{1}$$

where L_0 denotes the intrinsic luminance of the pavement and L_f the atmospheric luminance. In a foggy image, the intensity I of a pixel is the result of the camera response function crf applied to (1). Assuming that crf is linear, (1) becomes:

$$I = \text{crf}(L) = Re^{-kd} + A_\infty(1 - e^{-kd}) \tag{2}$$

where R is the intrinsic intensity of the pixel, i.e. the intensity corresponding to the intrinsic luminance value of the corresponding scene point and A_∞ is the background sky intensity.

4. Detection and characterization of safety-critical visibility ranges

The first system is able to detect daytime fog and estimate safety-critical visibility ranges. To be run, this system only needs an accurate geometrical calibration of the camera with respect to the road plane.

4.1. Camera modelling

Assuming that the road is locally planar, the distance of a point located at the range d on the roadway can be expressed in the image plane, assuming a pinhole camera model [15]:

$$d = \frac{\lambda}{(v - v_h)} \tag{3}$$

where $\lambda = \frac{H\alpha}{\cos(\theta)}$ and $v_h = v_0 - \alpha \tan(\theta)$. H and θ respectively denote the mounting height and the pitch angle of the camera. $\alpha = \frac{f}{t_p}$ is an intrinsic parameter of the camera based its focal length f and the size t_p of a pixel. v_0 and v_h respectively denote the vertical position in the image of the projection center and of the horizon line (see Fig. 1).

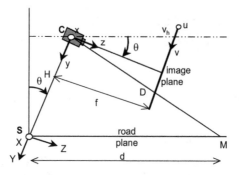

Figure 1. Modelling of the camera within the road environment. v_h: image line corresponding to the horizon line in the image.

The relationship between the distance d on the ground and the distance D from the same point M to the camera is as following:

$$D = \sqrt{H^2 \cos^2 \theta + (d - H \sin \theta)^2} \qquad (4)$$

4.2. Daytime fog detection

Following a change of d according to v based on (3), (2) then becomes:

$$I(v) = R - (R - A_\infty)\left(1 - e^{-\frac{k\lambda}{v - v_h}}\right) \qquad (5)$$

By taking the second derivative of I with respect to v, one obtains the following:

$$\frac{\partial^2 I}{\partial v^2}(v) = k\varphi(v)e^{-\frac{k\lambda}{v - v_h}}\left(\frac{k\lambda}{v - v_h} - 2\right) \qquad (6)$$

where $\varphi(v) = \frac{\lambda(R - A_\infty)}{(v - v_h)^3}$. The equation $\frac{\partial^2 I}{\partial v^2} = 0$ has two solutions. The solution $k = 0$ is not physically plausible. The only useful solution is (7):

$$k = \frac{2(v_i - v_h)}{\lambda} \qquad (7)$$

where v_i denotes the position of the inflection point of $I(v)$. In this manner, if $v_i > v_h$, daytime fog is detected and the parameter k is obtained. We deduce $V_{met} = \frac{3}{k}$ [11]:

$$V_{met} = \frac{3\lambda}{2(v_i - v_h)} \qquad (8)$$

To implement this principle, a region within the image that displays minimal line-to-line gradient variation when browsed from bottom to top is identified by a segmentation algorithm. A vertical band is then selected in the detected area. Finally, taking the median intensity of each segment yields the vertical variation of the intensity of the image and the position of the inflection point. Details of the method are given in [15]. It has been applied to a sample image in Fig. 2(a). Even if there are many vehicles in the original image, the method is able to ignore them and to detect fog presence, as well as to estimate the meteorological visibility.

4.3. Estimation of the visibility distance

The previous method detects that the visibility is reduced by daytime fog and estimates its density. In the same way, methods dedicated to other meteorological phenomena quantification could be added. Nevertheless, to supervise these different methods, a generic method is needed to estimate the visibility. To achieve this aim, we proposed to compute the distance to the furthest visible object on the road surface. This distance is called the mobilized visibility distance V_{mob}, which is close to the definition of V_{met} if a 5% contrast threshold is chosen [16]. Thus, a local contrast computation algorithm, based on Köhler's binarization technique and described in detail in [16], is applied to the image to compute local contrasts above or equal to 5%. The obtained contrast map contains objects of the road scene. A flat road may be assumed. As a matter of fact, along a top-bottom scanning line of the local contrast map starting from the horizon line, objects encountered get closer to the camera. Consequently, the algorithm consists of finding the highest point in the contrast map having a local contrast above 5%. v_c denotes the corresponding image-line. The distance to this point can then be recovered using Eq. (3), which allows estimating V_{mob} [17]:

$$V_{\text{mob}} = \frac{\lambda}{v_c - v_h} \tag{9}$$

However, the image may also contain vertical objects, which do not respect the flat world assumption and alter the method. This scenario is the case in Fig. 2(b), where the vehicle lights are detected higher in the image than the road surface elements. Another step is thus needed to filter the vertical objects and correctly estimate the visibility distance. This task is achieved using a background modelling method [17].

4.4. Camera specification

First, according to the sensor requirements given in section 3.1, the visibility system shall detect visibility up to d_{max} (400 m in our case). By using Eq. (3), the surface covered by a pixel at the distance d can be computed [15]:

$$\Delta(d) = \frac{\lambda}{\lfloor v_h + \frac{\lambda}{d} \rfloor - v_h} - \frac{\lambda}{\lceil v_h + \frac{\lambda}{d} \rceil - v_h} \tag{10}$$

where $\lfloor x \rfloor$ designates the whole part of x and $\lceil x \rceil$ the integer greater than or equal to x. We proposed this surface to be lower than 10% of d_{max} (40 m in our case), which is a good

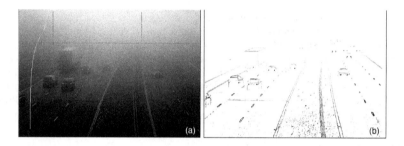

Figure 2. Detection algorithms applied to a fog highway image: (a) The vertical yellow curve represents the instantiation of (2); the horizontal red line represents the estimation of the visibility distance. The blue vertical segments represent the limits of the vertical band analyzed. (b) Map of local contrasts above 5%.

compromise between accuracy and cost [18]:

$$\Delta(d_{\max}) < 0.1 d_{\max} \tag{11}$$

Second, the system must detect fog. Based on section 4.2, the horizon line must lie in the image. Third, the visibility system shall detect visibilities lower than $d_{\min}$ (50 m in our case). To run correctly, the corresponding location of the inflection point must lie in the upper part of the image, i.e. v_i must be lower than v_0. Consequently, additional constraints on the sensor are as following:

$$v_h > 0 \tag{12}$$

$$v_h + \frac{3\lambda}{d_{\min}} < v_0 \tag{13}$$

From (12) and (13), the following inequation is obtained:

$$\sin^{-1}\left(\frac{H}{3d_{min}}\right) < \theta < \tan^{-1}\left(\frac{v_0}{\alpha}\right) \tag{14}$$

The admissible solutions of Eq. (14) can then be used to solve Eq. (11). To fulfill the requirements expressed in [13], we are thus able to specify relevant camera characteristics, which are partly detailed in Tab. 2. In this table, b denotes the diameter of the camera matrix. H denotes the sensor mounting height. f denotes the focal length of the camera optics (see Fig. 1). t_p denotes the pixel size of the camera matrix and dim_y the width of the matrix. θ denotes the pitch angle of the camera.

b [inch]	H [m]	f [mm]	t_p [μm]	dim_y [pix]	θ [degree]
1/2	6	4.5	4.65	1360	28-29

Table 2. Technical solution for camera specification.

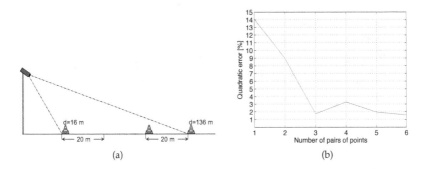

(a) (b)

Figure 3. Experimental verification of camera specifications: (a) experimental setup; (b) quadratic error of calibration with respect to the number of considered pairs of points.

4.5. Experimental validation

4.5.1. Verification of Camera Specifications

First, an experimental verification of the camera specifications has been carried out to check if we are able to reach the specifications. To achieve this aim, seven cones were set on a flat road section following the experimental setup of Fig. 3(a). Using the perspective projection model and the positions of the different cones, we are able to calibrate the camera:

$$(\lambda, v_h) = \underset{n=1..7}{\mathrm{argmin}} \sqrt{\sum_{i=1}^{n} \left(d_i - \frac{\lambda}{v_i - v_h} \right)^2} \tag{15}$$

where $\lambda = \dfrac{d_1 - d_n}{\frac{1}{v_1 - v_h} - \frac{1}{v_n - v_h}}$. The quadratic error is plotted in Fig. 3(b) with respect to the number of pairs of points taken into account in the calibration process. Three pairs of points, i.e. four points on the ground, suffice to obtain a quadratic error which is smaller than 2% at $d = 136$ m. This error is in line with the theoretical error at the same distance and hence acceptable.

4.5.2. Implementation of the System

The complete image acquisition system has been installed in a van equipped with a pneumatic pole. Using this vehicle, we have grabbed a sequence of images during sunrise. Sample pictures of this fog episode are shown in Fig. 4(a). The visibility distance has been estimated and is plotted in Fig. 4(b) with respect to the time. As one can see, the visibility distance increases as well as the global illumination in the scene while fog dissipates. The behavior of the system is good except at time $t \approx 50$ min, where the visibility is underestimated. This underestimation is due to the fact that the exposure time is momentarily too high so that the images are overexposed and contrasts deteriorated. Fortunately, the auto-exposure algorithm quickly solves the problem and the visibility increases again up to the maximum value, which is above 400 m as expected.

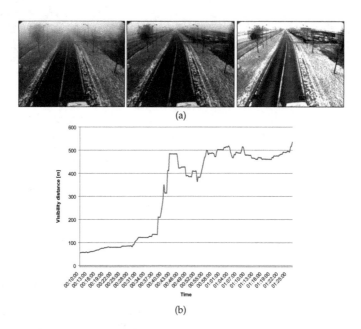

(a)

(b)

Figure 4. Implementation of the designed camera-based visibility metering system: (a) sample images acquired during a fog episode; (b) visibility distance estimated during this fog episode.

5. Monitoring of meteorological visibility

With the second system, we are able to monitor a whole spectrum of visibility ranges (from 0 to 10,000 m). In this system, we calibrated a response function of the contrast within the image with respect to reference visibility measurements obtained by external sensors. The camera needs not be a high resolution one. However, the calibration is more complex and needs at least one fog episode.

5.1. Contrast of a distant target

Let us consider an outdoor scene where targets are distributed continuously at increasing distances from the camera. When we assume that the surface of the targets is Lambertian, the luminance L at each point i of the target is given by:

$$L = \rho_i \frac{E}{\pi} \tag{16}$$

where E denotes the global illumination and ρ_i denotes the albedo at i. Moreover, it is a classical assumption to set $L_\infty = \frac{E}{\pi}$ so that the contrast of two Lambertian targets at distance d becomes [19]:

$$C = (\rho_2 - \rho_1)e^{-\beta d} \approx (\rho_2 - \rho_1)e^{-\frac{3d}{V}} = \Delta\rho e^{-\frac{3d}{V}} \tag{17}$$

Consequently, the contrast of a distant Lambertian target only depends on its physical properties and on its distance to the sensor and on the meteorological visibility distance, and no longer on the illumination. Such targets allow for computing contrasts in the scene in a way which is robust to strong variations in illumination [19].

5.2. Probabilistic modelling

Let us denote ϕ the probability density function (p.d.f.) of observing a contrast C in the scene:

$$\mathbb{P}(C < X \leq C + dC) = \phi(C)dC \tag{18}$$

The expectation of the contrast m in the image is expressed as [19]:

$$m = \mathbb{E}[C] = \int_0^1 C\phi(C)dC \tag{19}$$

Based on (17), C is a random variable which depends of the two random variables d and $\Delta\rho$. These two variables are assumed to be independent, which allows expressing Eq. (19) as [19]:

$$m = \mathbb{E}\left[\Delta\rho\right]\mathbb{E}\left[e^{-\frac{3d}{V}}\right] = \overline{\Delta\rho}\int_0^{+\infty}\psi(d)e^{-\frac{3d}{V}}dd \tag{20}$$

where $\overline{\Delta\rho}$ denotes the mean albedo difference between the objects in the scene and ψ denotes the p.d.f. of there being an object at the distance d in the scene. Choosing a suitable target distribution ψ allows us computing the expectation of the contrast using Eq. (20) with respect to the meteorological visibility distance V.

5.3. Expectation of the mean contrast

In this paragraph, we seek an analytical expression of Eq. (20). To achieve this aim, we assume a scene which contains n Lambertian targets with random albedos located at random distances between 0 and d_{max}. For a given sample scene, we can compute the mean contrast of the targets with respect to the meteorological visibility distance and plot the corresponding curve. Sample curves are plotted in blue in Fig. 5 ($n = 100$ and $d_{max} = 1000$ m). We can compute the mathematical expectation of the mean contrast and obtain the following analytical model:

$$m_u = \frac{V\Delta\bar{\rho}}{6d_{max}}\left[1 - \exp\left(-\frac{3d_{max}}{V}\right)\right] \tag{21}$$

where $\Delta\bar{\rho}$ is the mean albedo difference of the targets in the scene. We plot this model in red in Fig. 5. When we do not have any a priori on the targets distribution in the scene, this analytical model is the most probable with which to fit the data [19]. This fact is experimentally assessed in section 5.5. At this stage, we can make a comparison with the charging/discharging of a capacitor. The capacitance of the system is determined by the distribution of Lambertian targets in the scene. The smaller the capacitance of the system is,

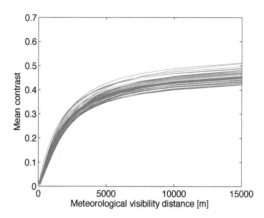

Figure 5. Blue: curves depicting the mean contrast in random scenes with respect to the meteorological visibility distance. Red: expectation of the mean contrast.

the faster the curves reach a 0.5 contrast. We thus define an indicator τ of the system quality which is the meteorological visibility distance at which two thirds of the "capacitance" is reached. A high value of τ also means a lower sensitivity of the model at low meteorological visibility distances.

5.4. Model inversion and error estimation

In the previous section, we have computed an analytical expression of the mean contrast expectation m_u with respect to the meteorological visibility distance V. Ultimately, we would like to compute V as a function of m_u. To achieve this aim, we need to invert the mean contrast expectation function (21). The inversion of this model exists and is expressed by [19]:

$$V(m_u) = \frac{3m_u d_{\max}}{1 + m_u W\left(\dfrac{e^{\frac{-1}{m_u}}}{m_u}\right)} \tag{22}$$

where the Lambert W function is a transcendental function defined by solutions of $W(x)e^{W(x)} = x$ [20].

5.5. Experimental evaluation

In this section, we present an experimental evaluation of the proposed model for visibility estimation. To achieve this aim, we have collected ground truth data.

5.5.1. Methodology

5.5.1.1. Instrumentation

The observation field test we used is equipped with a reference transmissometer (Degreane Horizon TI8510). It serves to calibrate different scatterometers (Degreane Horizon DF320)

(a)   (b)

Figure 6. Instrumentation of our observation field test: (a) the camera grabbing pictures of the field test;(b) the scatterometer along with the background luminancemeter.

used to monitor the meteorological visibility distance in France, one of which provided our data. They are coupled with a background luminance sensor (Degreane Horizon LU320) which monitors ambient light conditions. We have added a camera which grabs images of the field test every ten minutes. The camera is an 8-bit CCD camera (640×480 definition, H=8.3 m, $\theta = 9.8^o$, $f_l = 4$ mm and $t_{pix} = 9 \ \mu m$). Compared to the camera specified in section 4.4, it is thus a low cost camera which is representative of common video surveillance roadside cameras (cf. section 2.2). Fig. 6(a) shows the installed camera and its orientation with respect to the field test. Fig. 6(b) shows the scatterometer and the background luminancemeter.

5.5.1.2. Data collection.

We have collected two fog events the 28th February and the 1st March 2009. The fog occurred early in the morning and lasted a few hours after sunrise. During the same days, there were strong sunny weather periods. Fig. 7 shows sample images of sunny weather and cloudy weather and foggy weather. The corresponding meteorological visibility distances and luminances are plotted in Figs. 7(d,e). Obviously, the meteorological visibility distance ranges from 100 m to 35.000 m and the luminance ranges from 0 to 6.000 cd.m^{-2}.

We have thus collected exceptional experimental data. Indeed, we met rapidly changing weather conditions over a short period of time. The ranges of meteorological visibility distance and luminance were very large. In the literature, works are dedicated to limited ranges of visibility distances [17, 21]. For example, road safety applications are dealing with 0-400 m [17] whereas people working on environmental issues are dealing with meteorological visibility distances which are above 1000 m [21]. We are among the first to have collected data encompassing both ranges. Moreover, since the data was collected over a short period of time, we consider that the content of the scene did not change. For example, we assumed that the phenology of the trees did not change, so that the amount of texture in the scene without fog remains constant.

5.5.1.3. Location of Lambertian surfaces

To estimate m_u and thus V, we compute the normalized gradient only on the Lambertian surfaces of the scene as proposed in section 5.1. We thus need to locate Lambertian surfaces in the images. To achieve this aim, we compute the Pearson coefficient, denoted $P_{i,j}^L$, between the intensity of pixels in image series where the position of the sun changes and the value of the background luminance estimated by the luminancemeter. The closer $P_{i,j}^L$ is to 1, the stronger

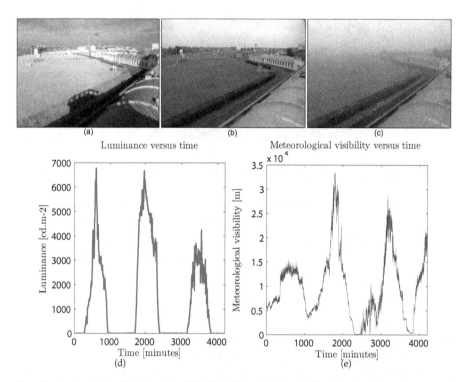

Figure 7. Samples of data collected in winter 2008-2009: (a) images with strong illumination conditions and presence of shadows; (b) cloudy conditions; (c) foggy weather situation; (d) background luminance and (e) meteorological visibility distance data collected in the field test during two days.

Figure 8. Mask of Lambertian surfaces on our field test: The redder the pixel, the higher the confidence that the surface is Lambertian.

the probability that the pixel belongs to a Lambertian surface. This technique provides an efficient way to locate some Lambertian surfaces in the scene. For our field test, the mask of Lambertian surfaces is shown in Fig.8. The redder the pixel, the more the surface is likely to be Lambertian.

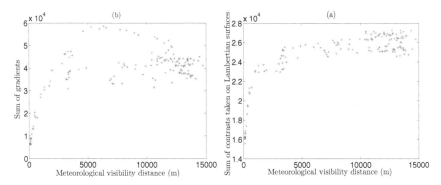

Figure 9. Visibility estimators: (a) the estimator is based on the contrast on Sobel gradient alone; (b) the estimator is based on Sobel gradient weighted by Lambertian surfaces.

Having located the Lambertian surfaces, we can compute the gradients in the scene by means of the module of the Sobel filter. For each pixel, we normalize the gradient $G_{i,j}$ by the intensity of the background. Since our camera is equipped with an auto-iris, the background intensity A_∞ is most of the time equal to $2^8 - 1$, so that this step can be avoided. Each gradient is then weighted by $P_{i,j}^L$, the probability of a pixel to belong to a Lambertian surface where no depth discontinuity exists (P^L is mostly very small). Consequently, only relevant areas of the image are used, and the scene need not be totally Lambertian. Finally, the estimated contrast in the scene $\tilde{m}_u$ is given by [19]:

$$\tilde{m}_u = \sum_{i=0}^{h} \sum_{j=0}^{w} \Delta\rho_{i,j} \exp\left(-\frac{3d_{i,j}}{V}\right) \approx \sum_{i=0}^{h} \sum_{j=0}^{w} \frac{G_{i,j}}{A_\infty} P_{i,j}^L \tag{23}$$

where $\Delta\rho_{i,j}$ is the intrinsic contrast of a pixel in Eq. (17), and h and w are respectively the height and the width of the images.

5.5.2. Results

5.5.2.1. Contrast estimators

We have computed Eq. (23) for our collection of 150 images with different meteorological visibility distances. For comparison purposes, we have also computed the simple sum of gradients in the image without weighting the Lambertian surfaces. The results are shown in Fig. 9. By using the Lambertian surfaces, we can see that the shape of the distribution in Fig. 9(a) looks like the curve proposed in Fig. 5, which is very satisfactory. Conversely, when all the pixels of the scene are used, the points are more scattered when the meteorological visibility distance is above 2500 m (see Fig. 9(b)): When the sky is clear and the visibility is high, the illumination from the sun strongly influences the gradients in the scene. Consequently, the estimation of the visibility is altered. These two distributions show the benefit of selecting the Lambertian surfaces to estimate the visibility distance.

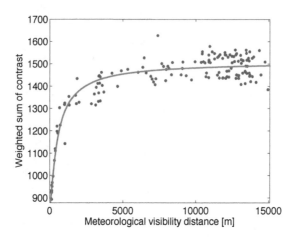

Figure 10. Data fitting with the mean contrast model. Dots: data. Red curve: fitted model.

5.5.2.2. Model fitting

We have to fit the mean contrast model (Eq. (21)) to the data shown in Fig. 9(a) using robust regression techniques. To ensure a mathematical solution, we have fitted the model (Eq. (24)), which is slightly different from the theoretical model. Three unknown variables a, b and d_{max} have to be estimated, which can be easily done using classical curve fitting tools.

$$\tilde{m}_u = \frac{aV}{d_{\max}}\left[1 - \exp\left(-\frac{3d_{\max}}{V}\right)\right] + b \qquad (24)$$

This model fits well with the data ($R^2 = 0.91$). In particular, we obtain $d_{\max} = 307.2$ m. The fitted curve is plotted in Fig. 10. We estimated a capacitance (as defined in section 5.3) of the system $\tau \approx 950$ m.

5.5.2.3. Discussions

From the fitted model, we can now invert the model using (22) and estimate the meteorological visibility distance $\tilde{V}$ based on the mean contrast m_u[19]:

$$\tilde{V} = \frac{3d_{\max}(b - m_u)}{(b - m_u)W\left(\dfrac{ae^{\frac{a}{b-m_u}}}{b - m_u}\right) - a} \qquad (25)$$

After having estimated the meteorological visibility distance, we can compute the error on this estimation. The results are given in Tab. 3. Since the applications are very different depending on the range of meteorological visibility distances, we have computed the error and the standard deviation for various applications: road safety, meteorological observation and air quality. One can see that the error remains low for critical safety applications. It increases for higher ranges of visibility distances, and becomes huge for visibility distances above 7 km. Different issues may be discussed. First, the model presented in this chapter is relevant for uniform distributions of distances which happen in many environments, such as highway scenes. The scene in which the experimental data used in this paper were collected may be meet this assumption. Second, the Sobel operator is certainly not the best estimate for the gradient. Indeed, it is a simple high-pass filter which is problematic because of the impulse noise of camera sensors. Different filters may be used to enhance the images beforehand, or to compute the contrast more robustly.

6. Outcome

First, a camera-based system has been developed to detect safety-critical visibility conditions. Sample results based on data collected during a morning fog episode are shown in the Fig. 4. When visibility is below 400 m, the accuracy of the system is expected to be 90%. The second system has been assessed on the test site of Météo France in Trappes, where images have been grabbed with visibility data (see Fig. 7). Based on our experiments, we are able to obtained error estimates which are lower than 10% [19] as well.

We thus have at disposal two different techniques to estimate the visibility distance in case of fog or haze. The first technique is dedicated to low visibility ranges but needs a high-resolution camera along with a simple calibration process. The second technique is not restricted to low visibility ranges and works with a low cost camera but needs a more complex calibration procedure. It is clear that the complementarity between both approaches must be studied so as to build a single system, which is easy to set up and deploy.

7. Potential applications

7.1. Winter maintenance

The monitoring of meteorological visibility has different applications for winter maintenance [22]. First, the knowledge of a low visibility area is important for the safety of winter maintenance operations. Second, a sudden drop of the visibility can be due to heavy snow falls. The relationship between liquid equivalent snowfall rate and visibility has also been investigated [23], which means that a camera-based visibility meter is potentially a good snow sensor. Finally, meteorological models have been developed to forecast pavement temperatures as well as snow height [24]. Nebulosity and fog are phenomena which alter the prediction, since the radiative transfer between the pavement and the air is affected. The assimilation of visibility data in these prediction models may be useful to increase the accuracy of forecasts.

Application	Highway fog	Meteorological fog	Haze	Air quality
Range [m]	0-400	0-1000	0-5000	0-15000
Number of data	13	19	45	150
Mean error [%]	12.6	18.1	29.7	-
Std [%]	13.7	18.9	22	-

Table 3. Relative errors of meteorological visibility distance estimation with respect to the envisioned application.

7.2. Fog nowcasting

The forecasting of the weather within the next six hours is often referred to as nowcasting. In this time range, it is possible to forecast smaller features such as individual showers and thunderstorms with reasonable accuracy, as well as other features too small to be resolved by a computer model. [25] show that combining satellite-based forecasting of low clouds with terrestrial measurements of humidity allows computing a probability of fog occurrence. A camera-based visibility meter could easily substitute for a humidity sensor. Indeed, using camera-based visibility estimation and meteorological data, [26] showed that visibility can be predicted up to 15 minutes in advance with 1-km mesh meteorological data. Such camera-based nowcasting methods may be good solutions to allow the re-routing of vehicles before they reach a low-visibility area in a timely manner.

7.3. Pile-up prevention and mitigation

[27] has proposed a review of best practices in terms of mitigation of highway visibility problems, in particular fog related issues. In this paper, he describes existing installations in the USA dedicated to driver alert in case of low visibility on the highway. Apart from fog dispersal techniques, the best practices are related to the timely alert of the drivers which approach a foggy area. Then, depending on the fog density, different advisory speed limits may be posted. In the same time, the public lighting is adapted. The component of these systems are made of weather stations, CCTV cameras and Variable Message Signs (VMS). More recently, Caltrans has installed the "Fog Pilot" system in District 6, which provides a high-technology solution every $\frac{1}{4}$ mile along a 12-mile (20-km) portion of State Route 99. This centralized solution relies on the use of infrastructure to vehicle communications to warn the drivers whose vehicles are equipped with a receiver, in case of sudden low speed area. Thanks to the proposed camera-based visibility monitoring techniques, we are able to build a decentralized fog-pilot, which makes use of CCTV cameras to monitor the visibility and allows optimizing the speed of drivers approaching a low visibility area, as well as the intensity of road studs. Based on best existing practices, its principle is to warn the drivers of a foggy area with enough time, so that they can adapt their speed to the prevailing visibility distance in the dangerous area.

8. Conclusion and perspectives

Reliable solutions to accurately monitor the meteorological visibility along road networks at reasonable costs are still not available. The use of the cameras, which are multifunctional sensors and are already deployed along the roadsides, is a promising solution. However, progress in computer vision are still needed to obtain robust techniques, which are able to fulfill the needs of transportation safety, of meteorology and of environment in terms of observation.

In this chapter, we have presented two different camera-based systems to estimate the visibility distance. These systems could be integrated in ITS stations to alert drivers as well public authorities in case of fog hazard. The first system is dedicated to safety-critical visibility ranges but needs a high-resolution camera along with a simple calibration process. The second technique is not restricted to low visibility ranges and needs a low cost camera but a more complex calibration procedure. In the future, it is obvious that the complementarity between both approaches must be studied so as to build a single system, which will be easy to set up and to deploy. Second, the problem of night fog is still subject to research. In particular, we would like to adapt our in-vehicle techniques of night fog detection to visual surveillance. Third, we would like to develop an intelligent camera where fog detection is implemented along with existing traffic applications (virtual loops, AID, etc.). Finally, we would like to develop test beds for our sensing technologies and our decentralized fog-pilot.

Author details

Nicolas Hautière[1], Raouf Babari[1], Eric Dumont[1],
Jacques Parent Du Chatelet[2] and Nicolas Paparoditis[3]

1 Université Paris-Est, Institut Français des Sciences et Technologies des Transports, de l'Aménagement et des Réseaux, France
2 Météo France, France
3 Université Paris-Est, Institut Géographique National, France

References

[1] M. Abdel-Aty, A.-A. Ekram, H. Huang, and K. Choi. A study on crashes related to visibility obstruction due to fog and smoke. *Accident Analysis and Prevention*, 43:1730–1737, 2011.

[2] B. Whiffen, P. Delannoy, and S.. Siok. Fog: Impact on road transportation and mitigation options. In *National Highway Visibility Conference, Madison, Wisconsin, USA*, May 2004.

[3] J. Kang, R. Ni, and G. J. Andersen. Effects of reduced visibility from fog on car-following performance. *Transportation Research Record: Journal of the Transportation Research Board*, (2069):9–15, 2008.

[4] F.D. Shepard. *Reduced Visibility Due to Fog on the Highway*. Number 228. 1996.

[5] C. A. Mac Carley. Methods and metrics for evaluation of an automated real-time driver warning system. *Transportation Research Record: Journal of the Transportation Research Board*, (1937):87–95, 2005.

[6] T. Pejovic, V. A. Williams, R. B. Noland, and R. Toumi. Factors affecting the frequency and severity of airport weather delays and the implications of climate change for future delays. *Transportation Research Record: Journal of the Transportation Research Board*, (2139):97–106., 2009.

[7] A. H. Perry and L. J. Symons. *Highway Meteorology*. University of Wales Swansea, Swansea, Wales, United Kingdom, 1991.

[8] N. Jacobs, Burgin W., N. Fridrich, A. Abrams, K. Miskell, B. Brswell, A. Richardson, and R. Pless. The global network of outdoor webcams: Properties and apllications. In *ACM International Conference on Advances in Geographic Information Systems*, 2009.

[9] R. Hallowell, M. Matthews, and P. Pisano. An automated visibility detection algorithm utilizing camera imagery. In *23rd Conference on Interactive Information and Processing Systems for Meteorology, Oceanography, and Hydrology (IIPS), San Antonio, TX, Amer. Meteor. Soc.*, 2007.

[10] S. Roquelaure, R. Tardif, S. Remy, and T Bergot. Skill of a ceiling and visibility local ensemble prediction system (leps) according to fog-type prediction at paris-charles de gaulle. *Airport. Weather and Forecasting*, 24:1511–1523, 2009.

[11] CIE. *International Lighting Vocabulary*. Number 17.4. 1987.

[12] WMO. *Guide to Meteorological Instruments and Methods of Observation*. Number 8. World Meteorological Organization, 2008.

[13] AFNOR. Road meteorology - gathering of meteorological and road data - terminology. NF P 99-320, April 1998.

[14] W.E.K. Middleton. *Vision through the atmosphere*. University of Toronto Press, 1952.

[15] N. Hautière, J.-P. Tarel, J. Lavenant, and D. Aubert. Automatic fog detection and estimation of visibility distance through use of an onboard camera. *Machine Vision Applications*, 17(1):8–20, 2006.

[16] N. Hautière, R. Labayrade, and D. Aubert. Real-time disparity contrast combination for onboard estimation of the visibility distance. *IEEE Transactions on Intelligent Transportation Systems*, 7(2):201–212, June 2006.

[17] N. Hautière, E. Bigorgne, and D. Aubert. Visibility range monitoring through use of a roadside camera. In *IEEE Intelligent Vehicles Symposium*, 2008.

[18] J.D. Crosby. Visibility sensor accuracy: what's realistic? In *12th Symposium on Meteorological Observations and Instrumentation*, 2003.

[19] N. Hautière, R. Babari, E. Dumont, R. Brémond, and N. Paparoditis. *Lecture Notes in Computer Science, Computer Vision - ACCV 2010*, volume 6495, chapter Estimating Meteorological Visibility using Cameras: A Probabilistic Model-Driven Approach, pages 243–254. Springer, March 2011.

[20] R. M Corless, G. H. Gonnet, D. E. G. Hare, D. J. Jeffrey, and D. E. Knuth. On the Lambert W function. *Advances in Computational Mathematics*, 5:329–359, 1996.

[21] C.-H. Luo, C.-Y. Wen, C.-S. Yuan, C.-C. Liaw, J.-L. ans Lo, and S.-H. Chiu. Investigation of urban atmospheric visibility by high-frequency extraction: Model development and field test. *Atmospheric Environment*, 39:2545–2552, 2005.

[22] Y. Nagata, T. Hagiwara, K. Araki, Y. Kaneda, and H. Sasaki. Application of road visibility information system to winter maintenance. *Transportation Research Records: Journal of the Transportation Research Board*, 2055:128–138, 2008.

[23] Jothiram Vivekanandan Jeffrey Cole Barry Myers Charles Masters Rasmussen, Roy M. The estimation of snowfall rate using visibility. *Journal of Applied Meteorology and Climatology*, 38(10):1542–1563, 1999.

[24] L. Bouilloud, E. Martin, F. Habets, A. Boone, P. Le Moigne, J. Livet, M. Marchetti, A. Foidart, L. Franchistéguy, S. Morel, J. Noilhan, and P. Pettré. Road surface condition forecasting in france. *Journal of Applied Meteorology and Climatology*, 48(12):2513–2527, 2009.

[25] V. Guidard and D. Tzanos. Analysis of fog probability from a combination of satellite and ground observation data. *Pure and Applied Geophysics*, 164:1207–1220, 2007.

[26] N. Yasuhiro, T. Hagiwara, K. Takitani, F. Kawamura, Y. Kaneda, and M. Sakai. Development of a visibility forecast model based on a road visibility information system (RVIS) and meteorological data. In *Transportation Research Board Annual Meeting, Washington DC, USA*, number Paper 11-2342, 2011.

[27] N. McLawhorn. Mitigating highway visibility problems. In *National Highway Visibility Conference*, 2004.

Chapter 3

Forecasting Weather in Croatia Using ALADIN Numerical Weather Prediction Model

Martina Tudor, Stjepan Ivatek-Šahdan,
Antiono Stanešić, Kristian Horvath and Alica Bajić

Additional information is available at the end of the chapter

1. Introduction

Numerical weather prediction (NWP) models are one of the factors contributing to the complex process of providing an accurate weather forecast. A number of global NWP models forecast weather over the whole Earth seven or more days in advance. These large scale global models do not provide high-resolution details that can be important for the weather we feel in a particular point (sensible weather). These details are often provided by limited area models (LAMs) that cover a particular area of interest as will be exemplified in this chapter. The Croatian Meteorological and Hydrological Service (CMHS) uses ALADIN (Aire Limitée Adaptation Dynamique développement InterNational, ALADIN International Team, 1997) limited area model for the operational weather forecast. This chapter describes the operational NWP aspects in CMHS and provides an insight into high impact weather phenomena that are typical for this region.

The quality of NWP model forecast depends on the NWP model that is used to compute the meteorological forecast fields. Model forecast should improve with the improvements in the model equations and parameterizations used to describe the atmospheric processes (Pielke 2002, Durran 1999, Werner et. al 1997). Another factor is the quality and availability of the measured atmospheric data used as input for the data assimilation procedure that creates the initial conditions for the model forecast. Limited area models also require the lateral boundary conditions data which are another factor contributing to the quality of the weather forecast. Both initial and lateral boundary conditions are usually taken from a larger scale and lower resolution NWP model covering larger geographical area. The initial conditions can be modified using a local initialization procedure.

The operational forecast suite in CMHS uses the initial lateral boundary conditions from the ARPEGE (Action de Recherche Petite Echelle Grande Echelle, http://www.cnrm.meteo.fr/ gmapdoc/meshtml/guide_ARP/arpege.html) global NWP model run operationally at Meteo-France. Another set of initial and boundary conditions available in CMHS is from the Integrated Forecast System (IFS, http://www.ecmwf.int/research/ifsdocs/CY36r1/ index.html) of the European Centre for Medium-Range Weather Forecast (ECMWF).

The files received from Meteo-France and ECMWF do not contain the global model fields covering the whole Earth on the native ARPEGE or IFS grid and model levels. To optimize the data transfer, the model data is interpolated to a limited-area Lambert projection grid in a resolution similar to the one of the global model in that area and it is covering a wider geographical area than the local LAM that uses it. Since ALADIN is a spectral model, initial and boundary conditions contain the meteorological data covering the whole model domain in the form of spectral coefficients, and not only the data on the lateral boundaries.

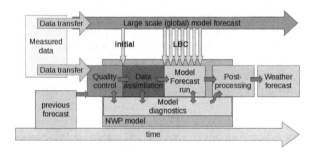

Figure 1. The scheme of the steps taken when making the weather forecast

The number of levels in the vertical is reduced (from 70 in ARPEGE or 91 in IFS to 37 in ALADIN Croatia), since LAM does not have to reach as high in the atmosphere as the global models. The global models have considerably longer forecast range (10 to 15 days in advance) that require knowledge of the atmospheric processes as high as the stratosphere top and mesosphere. These processes have negligible impact in the 72 hour forecast range used for the operational ALADIN forecast in CMHS..

The forecast process (Fig 1) consists of several steps:

1. collect the measured data representative for the initial forecast time from upper-air soundings, surface stations, meteorological satellites, aircraft and radar data,

2. data quality control which removes measurements if values are too far from the first guess, removal of observations according to empirical (black)list and surface observations with too large height difference between model orography and real station height.

3. establishment of the initial and lateral boundary conditions using data obtained from a larger scale (global) model,

4. the NWP model forecast run that consists of:

a. the data assimilation procedure that produces the analysis ("best possible" approximation of atmospheric initial conditions for the model forecast run),

b. the NWP model forecast,

c. diagnostics of the derived parameters from the model variables during the forecast run,

5. post-processing of the model output data.

Weather forecasts are time-critical applications. This means that the value of the products quickly degrades with time. In other words, the model forecast run has to finish before a certain time in the day, according to the requirements of the users. The choices of the input data and the model complexity that are used operationally also depend on the hour at which the weather forecast products should be available to the users. It is desired to keep the time span between the data measurements and the availability of the weather forecast as short as possible or within a reasonable interval usually determined by procedures in the forecast office or requirements of the other users. The time required for each forecast step in Figure 1 is determined by technical limitations such as the capacity of data transfer to the institution providing the weather forecast as well as the speed of the mainframe computer executing the NWP model software. The amount of data used in points (1) and (2) is limited by the capacity of data storages and speed of the data transfer lines. The complexity of the operational NWP model, as well as the size and grid resolution of the model domain is determined by this time span and the computer power. ARPEGE model data was the primary choice for the ALADIN operational model input in CMHS since it is available much earlier than the IFS model data (Fig 2). IFS is available later than ARPEGE because the data assimilation procedure starts later since it waits to assimilate more measured data. On the other hand, ARPEGE provides two sets of input data, as will be described later.

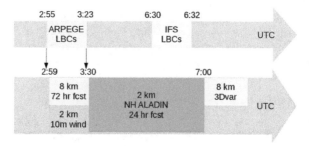

Figure 2. The schedule and timing of the 00 UTC operational run in August 2011.

A considerable effort in the research and development of the ALADIN model has been devoted to the model optimization and various model options with an aim to provide the best possible forecast in the short time using available computer power (Yessad, 2011). To improve the 10 m wind forecast in severe wind situations, like bura (local name for bora) windstorms, a procedure that provides a high resolution forecast of the wind 10 m above ground has been established. The high-resolution dynamical adaptation (HRDA) of the wind field using a hydrostatic version of the ALADIN model has been introduced to the operational suite soon after the 8 km resolution operational forecast runs have started in 2000 (Ivatek-Šahdan & Tudor, 2004).

2. Operational model characteristics

ALADIN has been developed by a group of scientists from 16 countries and shares a considerable part with the global models IFS and ARPEGE. Both global models are a result of a coordinated effort of ECMWF and Meteo-France. Another LAM developed on the basis of IFS global model is HIRLAM (High Resolution Limited Area Model, Unden et al 2002). These models share the same source code in many parts of the model. A considerable part of the dynamics and data assimilation is in common too. ALADIN can use the same vertical discretization, grid-point dynamics and the various physics options as the global model ARPEGE. This section describes the operational model characteristics, the domains and other options used in the operational suite.

2.1. Model domains used in the operational suite

The operational ALADIN model forecast is run on a Lambert-projection domain with 8 km horizontal resolution (Fig 3). The model fields are subsequently going through a dynamical adaptation procedure (Ivatek-Šahdan & Tudor, 2004) that produces 2 km resolution forecast of 10 m wind speed and gusts for a smaller domain shown in Figure 3. The procedure adapts the wind field of the 8 km resolution forecast and uses hydrostatic set-up of the ALADIN model with turbulence parameterization only. As an addition to the operational forecast, a 2 km resolution 24 hour forecast was established recently, that uses non-hydrostatic (NH) dynamics and the full parameterization set, including radiation, microphysics and convection schemes. This forecast is run for the same small domain (Fig 3).

2.2. Initialization and data assimilation

In the ALADIN community, the initial conditions for the model forecast can be obtained using different approaches:

1. using the large scale model data as initial conditions,

2. blending the large scale information from low resolution global model fields with the high resolution features from the previous LAM forecast,

3. using the data-assimilation procedure.

The first approach is used operationally in CMHS since the beginnings of the operational ALADIN forecast suite in 2000. There the upper-air and surface fields are taken from the

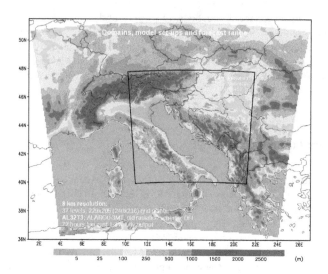

Figure 3. The ALADIN model domains and terrain height used operationally in CMHS.

ARPEGE global model and interpolated to the ALADIN grid. Afterwards, these fields are balanced using digital filter initialization (DFI) procedure (Lynch & Huang, 1994). However, the removal of the high-frequency wave energy from the initial conditions also affects the fast meteorological waves. Termonia (2008) showed how DFI can significantly reduce the depth of the eye of the storm and proposes a solution in the scale-selective DFI (SSDFI).

The second approach has not been used in CMHS, but blending has been used extensively in the ALADIN operational forecast suite in the Czech Meteorological and Hydrological Institute (CHMI, Brožkova et. al., 2001) as well as several other ALADIN member services (Hdidou, 2006). The blending procedure makes use of the fact that ALADIN is a spectral model. The initial model fields are constructed from the long waves coming from the low resolution model (ARPEGE) and short waves from the high resolution model (ALADIN).

The last approach is to initialize ALADIN model using data assimilation and currently it is used in experimental configuration in CMHS. Data assimilation is a procedure wherein information coming from measurements is combined with some a priori estimate (usually a short range forecast often called first-guess or background field) and their associated obser-vation and background model errors, respecting the model dynamical balances, to get the analysis - a maximal likelihood or "as best as possible" approximation of the true state of the atmosphere at a given time. There are different methods of solving this analysis problem (Hólm 2008) and at CMHS two of them are used: the optimal interpolation (OI, Courtier 1999) and the variational method (3DVAR, Hollingsworth et al, 1998). Optimal interpolation is used for updating land surface fields while 3DVAR is used for analysis of upper air fields. It was shown by Mahfouf (1991) that 2m analysis increments of temperature and humidity computed with OI can be used to update the land surface variables. As only the increments of 2m temperature

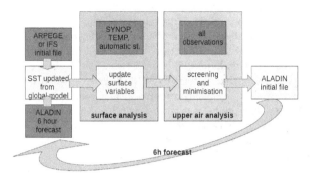

Figure 4. Schematic view of the data assimilation cycle.

and relative humidity are needed, and only the surface measurements are used, rather simple OI method was implemented for computing the 2m analysis increments. For the upper air fields more control variables are defined, more sources of data are used, and more sophisticated variational algorithm (3DVAR) is implemented. 3DVAR belongs to the group of intermittent variational data assimilation strategies, and whereas the true state of the atmosphere is never known, the principal challenge in determining the analysis is to estimate the background (model) errors. Namely, knowledge of the background error statistics is used to derive the analysis increments and transfer this information in horizontal and vertical directions in the atmosphere. The background error statistics is in CMHS estimated by using the so-called standard NMC method. In this approach, background error statistics is derived from a set of differences between forecasts that are valid at the same time, but at different forecast horizons, typically 12 and 36 hours. In addition, the multivariate formulation of the background error statistics that uses vorticity, divergence, temperature and surface pressure, and humidity as control variables (Berre, 2000), allows the propagation of the analysis increments from one to all the other variables, and thus increases the effect of each single observation. Finally, the potential use of the ensemble and seasonal formulations of the background error statistics is currently being investigated.

Practical implementation of data assimilation setup in CMHS is illustrated in Figure 4 that shows the scheme of one assimilation cycle. Because of limited computer resources the assimilation cycle is run in quasi-operational mode i.e. observation data is taken at the operational time, but the assimilation procedure is performed with certain time delay. An assimilation cycle is a sequence of a 6 hour forecast and analysis that is run on a daily basis. In the first step, SST in the 6h forecast from previous cycle (background) is replaced with SST taken from global model ARPEGE analysis. This replacement is done because no local SST analysis is implemented. In the second step, the background land surface variables are updated. As there are almost no or very little surface observations available, 2m observations are used to retrieve information about the surface temperature and soil water content. Therefore, after the quality control of the 2m observations, an optimal interpolation is used to calculate increments of 2m temperature and relative humidity. These increments in a more or less sophisticated way are propagated into the land surface equivalents and used for updating the land surface variables. In the next step, the background upper air fields are analysed using 3DVAR. The standard NMC background error statistics is calculated by the aforementioned procedure over a 100-day period, from 15 Feb – 25 May 2008. The background error matrix was not tuned a posteriori such as in some other NWP systems (Bölöni and Horvath, 2010). The 3DVAR procedure includes a quality control of available data (screening) and minimization procedure that as output produces an analysis which is used for initiating the 6h forecast. The assimilation cycle is repeated every 6 hours, at 00, 06, 12 and 18 UTC. It is also run with a time delay big enough to enable usage of long cut off ARPEGE coupling files (the ARPEGE model is run later and in assimilation all data is used) for boundary conditions of 6h ALADIN forecast. Also long cut off data is used in cycle (data available after the time period needed to collect all observations of interest). Production from the assimilation cycle is done following same steps as in the assimilation cycle but at its end, a 72h forecast is performed. However, to perform a quasi-operational 72 hours forecast, timing constrain does not allow to use long cut off data and long cut off ARPEGE files; thus short cut off data and ARPEGE files are used.

2.3. Model dynamics

The NWP model characteristics are usually described in the terms of the model numerics, dynamics and physics. Model numerics refers to the model computational domain, coordinate system, model resolution and grid as well as the mathematical methods used to solve the system of the prognostic differential equations.

Model dynamics refers to the resolvable processes that are resolved by the model grid and described by the set of model equations for horizontal and vertical momentum and conservation of mass and thermodynamic properties. These processes encompass advection, pressure gradient force and adiabatic changes of heat/temperature.

The model dynamics is computed using semi-implicit time integration scheme (Robert, 1982). An implicit treatment of the gravity wave equation is absolutely stable (Durran, 1999). The semi-implicit scheme treats implicitly only a linearized form of the adjustment terms in the shallow water equations. The method results in solving the Helmholtz equation in spectral space.

ALADIN is a shallow-water spectral limited-area model. It applies Fourier spectral representation of the model variables. That allows the advantage of fast Fourier transforms (FFTs) in both directions. An elliptic truncation that limits the Fourier series (Machenhauer & Haugen, 1987) ensures an isotropic horizontal resolution. The number of grid points in each horizontal direction of the whole integration area (N) is chosen so that N>3M+1, where M is the truncation wave-number in the same direction. This representation ensures that the nonlinear terms of the model equations are computed without aliasing. The model fields are transferred from spectral to grid-point space and back in each model time step.

The accumulation of energy at the shortest wavelengths, due to spectral blocking, is reduced by a common 4th order numerical diffusion at the end of the time step. Semi-Lagrangian interpolators can be more or less diffusive (Staniforth and Cote, 1991). Combining two interpolators of different diffusivity with the flow deformation as a weighting factor yields more physical horizontal diffusion scheme that is based on the physical properties of the flow (Vana et al., 2008). This semi-Lagrangian horizontal diffusion (SLHD) is combined with numerical diffusion that removes short waves from the high layers of the atmosphere.

The advection of the prognostic variables in the model is computed using two-time-level semi-Lagrangian scheme. The method takes the model grid-points as the arrival points of the trajectory. The trajectories are computed one time step backwards to the origin points.

The model prognostic variables that are involved in the semi-implicit computations in the hydrostatic version of the model are surface pressure, the horizontal wind components, temperature and water vapour. The non-hydrostatic dynamics involves two additional model variables: pressure departure and vertical divergence that are treated by the semi-implicit computations. The developments in physics have introduced more prognostic variables to the model, such as cloud water and ice, rain and snow, as well as convective updraft and downdraft vertical velocities and mesh fractions. These quantities can be (optionally) advected by the semi-Lagrangian scheme and diffused by SLHD but do not enter the semi-implicit computations. In the operational forecast run, all these variables are advected, but SLHD is applied only to water vapour, cloud water and ice. This configuration supports modelling the advection in the atmospheric front, but may spoil the forecast of rainfall due to orography or other local feature that does not move with the flow,

The finite difference method is used to solve the model equations in the vertical, on 37 levels of hybrid pressure type eta coordinate (Simmons & Burridge, 1981). The primitive prognostic equations are solved for the prognostic variables using the two time level, semi-implicit, semi-Lagrangian advection scheme with a second-order accurate treatment of the nonlinear residual (Gospodinov et al. 2001).

2.4. Lateral boundary conditions and coupling

LAM uses the large scale model data in a narrow coupling zone on the lateral boundaries and at discrete time intervals. The coupling of the model variables is done using Davies (1976) relaxation scheme in a narrow zone on the lateral boundaries of the LAM domain (Fig 5). The model dynamics requires usage of time dependent and periodic LBCs (Haugen & Machenhauer, 1993). The coupling procedure has to be applied at the very beginning or end of the gridpoint computations (Radnoti, 1995) due to constraints imposed by the model dynamics. Various schemes for the lateral boundary treatment are associated to different problems (Davies, 1983). Werner et al. (1997) give an overview of weaknesses of the LAM forecast caused by the LBCs.

The lateral boundary conditions (LBC) are operationally obtained with a 3 hour interval. This interval is a compromise between the need to reduce the amount of data that have to be transferred and stored and the need to capture fast cyclones, fronts and other meteorological phenomena that can enter the LAM domain through its lateral boundaries. The coupling scheme requires large scale data in the coupling zone for each LAM time step. The large scale model data in the coupling zone is interpolated linearly in time. The time span between the available LBC data of 3 hours and the coupling zone width is 8 grid-points in 8 km resolution. Three hours can be sufficiently long and 64 km can be narrow so that a particular meteorological feature can cross the coupling zone within the coupling interval (Fig 5). Solutions to this problem have been proposed, that include an alternative coupling scheme or interpolation in time (Tudor & Termonia, 2010), but have not been implemented to the operational model so far. The operational 2km resolution forecast uses LBC data from the operational 8 km resolution forecast with a 1 h interval.

In the two-way coupling approach, the flow information about the state of the atmosphere goes from the small-scale model to the large scale one, as well as in the opposite direction. This method requires simultaneous integration of both high and low resolution runs. For that reason, the operational 8km and 2 km resolution runs use one-way coupling that allows transfer of information only from the large-scale to the small-scale model run. The high resolution forecast starts only after the low resolution one is finished.

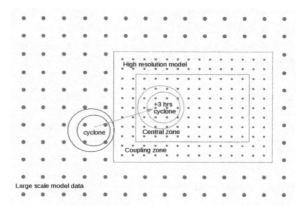

Figure 5. Schematic view illustrating the coupling problem.

2.5. Model physics

Model physics describe the processes unresolved by the model grid, radiation, cloud microphysics and surface processes. The processes described by the model physics are said to be parameterized (Pielke, 2002). The sub-grid-scale and diabatic processes are parameterized so their contribution to the changes in the large-scale state of the atmosphere can be considered.

The convective processes redistribute momentum, heat and moisture in the vertical. The deep convection parameterization (Fig 6) used in ALADIN (Gerard & Geleyn, 2005) is a prognostic mass-flux scheme (Gerard, 2007) where convective processes are treated with the use of prognostic variables for updraft and downdraft vertical velocities and mesh fractions (Gerard et al., 2009).

Cloud microphysics describe the processes of condensation, evaporation, freezing and melting as well as the processes that transform the cloud water droplets and ice crystals into rain and snow (Fig 6). ALADIN uses a simple microphysics scheme with prognostic cloud water and ice, rain and snow (Catry et al., 2007) and a statistical approach for sedimentation of precipitation (Geleyn et al., 2008). The microphysics scheme of the resolved processes is kept as close as possible to the original scheme that was without the prognostic condensates. The original scheme is the Kessler (1969) type scheme with modifications (Geleyn et. al. 1994) that include melting and freezing and imposing a brutal transition from ice to water at the temperature triple point.

The radiation processes described in the model encompass the transfer, scattering, absorption and reflection of the shortwave solar radiation and long-wave thermal radiation of the Earth's surface and clouds. There are several radiation schemes available in the ALADIN model (Morcrette, 1989, Mlawer et. al. 1997), but the simplest one is used in the operational version. The operational scheme (Ritter and Geleyn 1992) is based on Geleyn and Hollingsworth (1979) scheme. It is simple and computationally cheap since it uses only one spectral band for long-wave and one for short-wave radiation computations. The scheme has been enhanced recently (Geleyn et. al. 2005a, 2005b) but these modifications did not improve the cloud and temperature forecast in the stratus and fog case (Tudor, 2010).

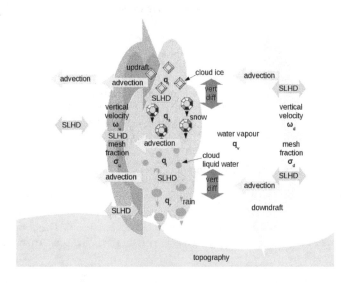

Figure 6. Schematic view of the scheme of the prognostic variables describing microphysics and convection and associated processes in the version of the ALADIN model used in CMHS.

The turbulence parametrization scheme describes the impact of the unresolved motion and surface roughness on the vertical transfer of heat, momentum and moisture. The turbulent exchange coefficients are modified from Louis et al. (1982) and are computed using prognostic values of TKE (turbulent kinetic energy) according to Geleyn et al. (2006) and Redelsperger et al. (2001).

The gravity wave drag and lift parameterization scheme describes the vertical momentum flux due to the atmospheric waves generated by the unresolved topographic features. The surface parameterization schemes describe the soil properties and their impact on the meteorological model variables. ISBA (Interaction Soil Biosphere Atmosphere) is the surface scheme used in the operational forecast (Noilhan & Planton, 1989) as well as in the surface data assimilation (Giard & Bazile, 2000). The scheme describes the exchange of heat and moisture between land surface and air using thermal and hydrological properties of particular soil types (described with dominant land-use type, percentages of clay, sand and silt, useful soil depth, thermal roughness length) and vegetation changes during the year (described by vegetation fraction, leaf-area index, surface resistance to evapotranspiration). The prognostic variables are temperature, liquid and solid water contents are computed on two layers representing the soil surface and deep soil properties, and additional variables describe surface snow reservoir, density and albedo and ice and water on leaves. Snow budget depends on snow precipitation, evaporation and melting as well as snow accumulated on vegetation.

Each parameterization scheme obtains the information on the state of the atmosphere from the model variables and possibly output from other parameterizations. Then it uses a set of closure assumptions that relate the parameterized process to the state of the atmosphere. The parameterization schemes have the largest impact on the prediction of the sensible weather at the Earth's surface (Pielke, 2002). These schemes became more complex over time and interact with each other, the numerical and dynamical parts of the model.

Model diagnostics produces the numerical values of the sensible weather fields from the model variables, such as the accumulated precipitation, instantaneous cloudiness or maximum wind gusts. Wind, temperature and humidity are interpolated from the model levels to the standard meteorological measurement heights (10 and 2 meters above surface) using a parameterized profile (Geleyn, 1988).

2.6. High resolution dynamical adaptation

The Croatian mountains are relatively small in extend, many of them have width close to 10 km and length less than 50 km. But the mountain peaks reach over 1 km inland, and several mountain tops reach over 1.5 km very close to the coastline (Fig 3). These mountains are separated by deep and inhabited valleys while important roads and connections go through the mountain passes.

Weather in Croatia is in many ways modified or even controlled by orography (Zaninović et. al. 2008), but the large scale models do not resolve many of these local weather patterns. These weather patterns result from interaction of synoptic forcing with orography. Improved representation of orography in a meteorological model is expected to improve the forecast of weather phenomena that are strongly influenced by local orography (Horvath et. al., 2011, Bajić, 2007, Branković et. al., 2008).

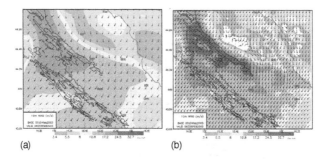

(a)   (b)

Figure 7. m wind forecast with 8 km (a) and 2 km resolution (b) for 6 UTC 5th Feb 2003.

The dynamical adaptation method of Žagar & Rakovec (1999) has been adapted for the purpose of operational forecast of the 10 m wind. The method provided successful operational forecast of the 10 m wind (Ivatek-Šahdan & Tudor, 2004) an has been used extensively in research impact studies (Bajić et al., 2007) as well as case studies (Tudor & Ivatek-Šahdan, 2002) of severe wind.

The meteorological model fields are first interpolated from the low resolution (8km in this case) to a higher resolution (2km) grid, but on considerably lower number of model levels, from 37 to 15 levels in the vertical. The number of vertical levels is reduced to minimize the computational cost. The levels close to the ground are of similar density, but the levels higher in the atmosphere are mostly omitted. Then a hydrostatic version of the ALADIN model is run for 30 time steps with a 60 second time step. The same large scale model data is used for the initial and LBC data, therefore the fields on the lateral boundaries do not change during the adaptation procedure. Turbulence is the only parameterization scheme used. Contributions from the moist and radiation processes are not computed to accelerate the model run.

HRDA improves 10 m wind forecast (by 15%), especially in weather situation with strong bura wind, since wind-speed and direction depend on the terrain configuration upstream. Figures 7a and 7b illustrates the impact of high resolution orography on the 10m wind forecast. HRDA wind is much weaker (reduces from 10-15 m/s to less than 5m/s) and changes the direction (up to 180°, depending on location) from the low resolution forecast in the valley upstream of the mountain. Downstream of the mountain, wind is much stronger (increases from 15 m/s in low resolution model to more than 30 m/s) in the high resolution model run as mountain wave breaks and windstorm reaches the mountain slope (Fig 8).

The 10 m wind forecast from HRDA has been used in several applications, and improved the safety of the traffic in the air, on the sea as well as on land. Unfortunately, there were also events of short bura episodes connected to a local pressure disturbance (Tudor & Ivatek-Šahdan, 2010) that were not predicted by HRDA. These were used in a further study to assess the impact of the neglected effects to these bura episodes. This procedure is able to forecast bura onset, duration and strength if it is a consequence of a synoptic forcing. It misses those cases initiated by small scale disturbances in the pressure fields. These disturbances have to be modelled first, using the non-hydrostatic model version on more levels in the vertical with

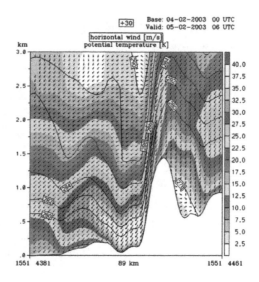

Figure 8. Vertical cross-section of wind speed (shaded), direction (vectors) and potential temperature (black isolines) for 6 UTC 5th Feb 2003.

full complexity of the physical parameterization package. These conclusions were based on analysis of a number cases of bura associated with a small scale pressure disturbance. Once a pressure disturbance is modelled, the wind field acts according to the measurements. The disturbances have to be detected first in the pressure measurements from the automatic stations that are performed on a less dense network than the wind measurements. The operational forecast using the latter configuration has been established only recently. The number of cases is still too short to make a thorough statistical analysis.

2.7. The operational forecast schedule

In CMHS, the operational ALADIN forecast is run twice per day, starting from 00 and 12 UTC analyses. The operational ALADIN model forecast is run 72 hours in advance on a Lambert-projection domain with 8 km horizontal resolution on 37 hybrid sigma-pressure levels in the vertical. The model fields are subsequently going through a dynamical adaptation procedure (Ivatek-Šahdan and Tudor, 2004) that produces 2 km resolution forecast of 10 m wind speed and gusts. There are two more sets of initial and LBC data available from both ARPEGE and IFS. The analyses at 06 and 18 UTC are used for cycling of the data-assimilation procedure. Only 6 hour forecasts are produced from these analyses and they are used as a first guess in the data assimilation procedure for the next analysis time (12 and 00 UTC) that initiates the 72 hour forecast. Alternative 8 km resolution 72 h forecasts run from the initial fields created in 00 and 12 UTC data-assimilation. These runs provide an alternative forecast fields that differs slightly from the first operational run. As an addition to the operational forecast, 2 km resolution 24 hour forecast was established recently, that uses the non-hydrostatic (NH) dynamics in the ALADIN model and the ALADIN's full parameterization set, including the convection scheme. This forecast runs is performed once per day, following the 00 UTC operational 8 km resolution forecast. It uses the 6 hour forecast from the 8 km resolution operational run as input initial file and runs with SSDFI. This high-resolution forecast is run for 24 hours, until 6 UTC on the next day. This procedure allows covering the 24 hour period used to collect precipitation data from the rain-gauges.

2.8. Predictability

High impact weather events are often forecasted by means of their predictability. ALADIN has been used in several predictability studies (e.g. Branković et al., 2007) as well as the Limited Area Ensemble Forecasting (Wang et al., 2011) system. In most of these studies, ECMWF ensemble forecasts were dynamically downscaled for several severe weather cases. Therefore, the initial conditions come from perturbations generated by singular vectors of ECMWF. Wang et al. (2011) combined different initial and boundary conditions from the perturbed global model with other aspects of forecast uncertainty, such as blending, multi-physics approach and breeding.

Predictability studies of bura cases have found little sensitivity to different initial conditions (Ivatek-Šahdan & Ivančan-Picek, 2006). Horvath et al. (2009) suggested a small effect of uncertainties in the upstream initial conditions on the bura events in the southern Adriatic Sea. Another study of severe bura (Branković et al., 2007) exemplified a case of gale force bura that was predicted with a probability exceeding 95%. Dynamical downscaling of ECMWF fields from 40 to 8 km resolution improves the precipitation rate and pattern in the intensive precipitation cases (Branković et al., 2008) too, the predicted precipitation is roughly double in the high resolution forecasts, but still reaches only 20% of the observed precipitation maximum.

3. High impact weather events

The capability of the ALADIN model to predict high-impact weather events in Croatia is illustrated by several exemplary cases that encompass various weather types.

3.1. Bura (bora)

Bura is a local word describing a downslope windstorm on the eastern Adriatic coast. It is a northeasterly wind with high gustiness that is mostly controlled by the upstream topography of Dinaric Alps. During a windstorm episode, several roads and ferry lines to the islands get closed. The strength and variability of bura wind varies from case to case (e.g. Horvath et al., 2009) and reaches up to 69 m/s and varies in space, although several places became famous for it. The processes that control bura occurrence and strength are briefly illustrated. Grisogono & Belušić (2009) give a comprehensive review of recent advances in bura research.

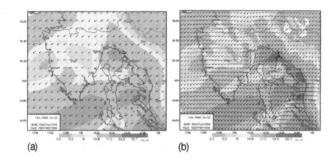

(a)   (b)

Figure 9. m wind forecast with 8 km (a) and 2 km resolution (b), for 15 UTC 7th Nov 1999.

MAP Special Observing Period 15 (SOP 15), from 7th to 9th Nov 1999, was characterized by severe bura on the eastern Adriatic Sea (Fig 9). The bura was associated with an intensive cyclone that was slowly moving over central Italy. A cold air outbreak from northeast intensified the pressure gradient over the coastal mountains and intensified wind in the lower layers of the atmosphere. During SOP 15, special measurements were taken from ELECTRA aircraft, through a northwest-southeast transect, parallel to the eastern Adriatic coast (Grubišić, 2004). The measurements were taken on levels 330 m and 660 m above the sea level and in the period from 15 to 16 UTC on 7th Nov 1999 (Fig 10). The 8km resolution forecast underestimated wind speed in the vicinity of the coastal mountains by 30% on average (Fig 9a). The wind forecast improved considerably (up to 10% over or under estimation of wind speed) using HRDA (Fig 9b). But the wind above the open sea (where the aircraft flew) was similar, both at surface (Fig 9) and at flight levels (Fig 10). The wind maximum at 660 m level is well predicted, as well as the northern minimum. The southern minimum is weaker by 5 m/s than measured and moved further south. On the lower level, at 330 m, model overestimates the maximum of the wind component normal to the flight direction by 3 m/s.

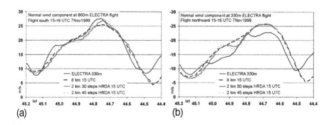

(a)   (b)

Figure 10. Comparison of measured and model normal wind components on flight levels 660 m (a) and 330 m, for 15-16 UTC 7th Nov 1999.

The strongest wind gust ever measured in Croatia was 69.0 m/s on Maslenica station (Bajić 2003) that was still operational during December 2003. During severe bura in that area, the 10 m wind speed is usually underestimated by the 8 km resolution forecast by 30 m/s on average

(Fig 11a). The HRDA forecast improves the surface wind forecast as it puts the wind speed maximum closer to the surface (Fig 11b), especially to the slope of the mountain as the mountain wave breaks. The wind maximum is still underestimated by 10-12%.

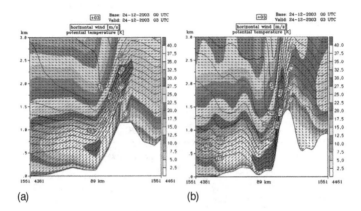

(a)   (b)

Figure 11. Vertical cross-section of wind speed (shaded), direction (vectors) and potential temperature (black isolines) from 8 km (a) and 2km (b) runs, for 3 UTC 24th Dec 2003.

There were several severe wind episodes in December 2003 (Fig 12), particularly in the period from December 22 to 26 2003, during which infrastructure was damaged on the highway going through the area hit by the severe bura, including the Baričević meteorological station (Fig 12a). The 2 km resolution forecast is in better agreement to the measured data for strong bura events (Fig 12a) as well as the most severe one (Fig 12b). In the absence of atmospheric soundings, these results confirm that the vertical structure of the meteorological fields in the lower troposphere also improves in high resolution for this case.

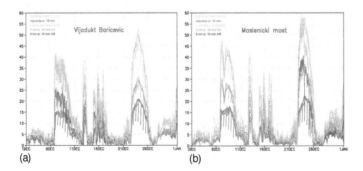

(a)   (b)

Figure 12. Comparison of measured 10 minute maximum (light blue) and average (blue), 8 km resolution mean wind forecast (red), 2 km resolution mean wind (orange) and wind gust forecast (yellow) at Baričević (a) and Maslenica (b) for December 2003.

3.2. Forecast of the road conditions

The bura variability in space and time has a pronounced influence on road traffic. Therefore, knowing bura characteristics is a necessary condition for road transport safety. To properly organize the traffic safety system, special emphasis should be given to the quality of measured long term wind speed and direction data and low resolution atmospheric forecast models.

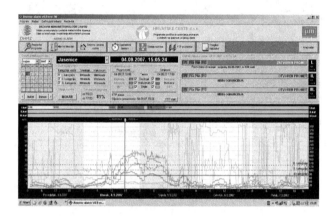

Figure 13. The ANEMO-ALARM user interface. The measured (thick lines) and modelled (thin line) mean wind speed is shown on a graph as blue line, wind gusts are red and direction is green.

An application, named ANEMO-ALARM (Bajić et. al., 2008), has been developed that assists road authorities in managing the traffic on the roads affected by strong wind and turbulence. The application is based on measured and forecasted wind speed and gusts for a choice of locations on Croatian roads that are most affected by severe wind. The application communicates with the user through a graphical user interface (Fig 13). The interface shows current and expected alarm status for road traffic safety conditions for any of the three categories of vehicles (green is for open road, yellow for preparedness status and expected road closure and red indicated that the road is closed).

3.3. Air crash investigation

Prognosed cloud water and ice, rain and snow mixing ratios are valuable output of the operational model forecast. Their three-dimensional distribution helps in identifying the areas that are a potential threat to aviation. This data can provide important information on the state of the atmosphere inside a cloud, especially when in-situ and remote measurements are sparse or insufficient to describe more detailed structure of the cloud. This value is briefly illustrated by means of a case when weather contributed to the crash of an airplane on the Velebit Mountain.

On 5th February 2009, a Cesna 303, crashed 500m north of the Vagan peak of the Velebit Mountain. The small airplane flew from Zagreb to Zadar, first at 2600 m, then at 2000 m feet.

Before the crash, the airplane descended from 1900 to 1600 m. The time of the accident was recorded as 13:53 UTC when air traffic control lost contact with the pilot.

Strong southwest wind prevailed throughout the troposphere. Wind speed increased with height, upstream of the mountain, in Zadar, the maximum wind speed of 20 knots was reached at 750m reaching 45 knots just below 3 km in the Zagreb radiosonde measurements. Temperatures on Zadar airport (88 m) and in Zagreb (128 m) were 14 to 15 °C. On Zavižan peak (1594 m) of the Velebit mountain temperature was 0.3°C, relative humidity was 100% and there was very low visibility, indicating that the mountain top was in a cloud. The satellite images show Velebit mountain covered by a thick cloud and mountain waves in the lee (Fig 14a). The cloud-top temperature was below 0°C (Fig 14b).

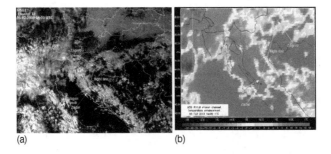

(a)   (b)

Figure 14. Satellite picture of clouds (a) and cloud top temperature (b) for 14 UTC 5th Feb 2009.

The model vertical cross-section of vertical velocity and potential temperature (Fig 15a) shows strong downward motion above and north of Vagan peak and variable vertical velocity downstream. Changes in the vertical velocity are accompanied by waves in the isolines of potential temperature (blue lines in Fig 15a). As the airplane flew from Zagreb to Zadar, it faced strong opposing wind and changing upward and downward motions that were stronger as the airplane approached the Vagan peak.

The mountain was covered by layers of clouds consisting of both cloud water and ice (Fig 15b) and high relative humidity existed in the cloud free area. The 0°C-isoline was above the 6500 feet flight level over the valley close to Zagreb, but the 0°C isoline descends below that height above the mountains close to Vagan peak (Fig 15b). Turbulent kinetic energy had high values above and downstream of the mountain (more than $2.5*10^4$ m^2/s^2).

3.4. Jugo (sirocco)

The Adriatic region is often affected by another severe wind that is rather familiar to the local population. Jugo is a local word describing the southeasterly wind in the eastern Adriatic. It interupts the ferry lines to the islands as well as other activities on the sea and the coast since it generates large sea waves due to the long fetch. It is often associated with heavy rainfall. The strength of jugo wind is often underestimated in southern Adriatic, especially in Dubrovnik

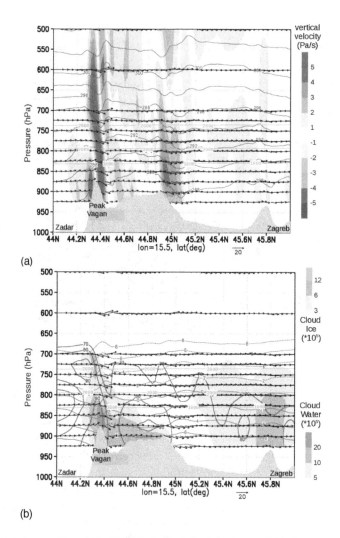

Figure 15. Vertical cross-sections of (a) vertical velocity (shaded), wind vectors parallel to the cross-section (vectors), potential temperature (K, blue lines) and (b) cloud water and ice (shaded), relative humidity (blue lines) and temperature (°C, red lines) for 14 UTC February 5, 2009. The flight levels are shown as horizontal green lines.

and on Palagruža, Lastovo and Vis islands by 15% (2 km resolution) to 30% (8km resolution) on average. The time-variability of jugo wind speed and direction is underestimated over the open sea (at least in some cases), as was revealed in the comparison with the measurements in the open sea during the DART (Dynamics of the Adriatic in Real-Time) oceanic research cruise in March 2006 (Tudor, 2011). The model error of the wind speed forecast is not uniform and can exceed 50% in some intervals for both 8 and 2 km resolution forecast.

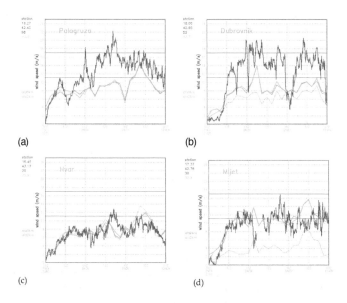

Figure 16. Comparison of measured (purple), 2 km (full lines) and 8 km resolution wind speed forecast (dashed) for Palagruža (a), Dubrovnik (b), Hvar (c) and Mljet (d) for 7th-9th Nov 2010.

3.5. Heavy rain

Large amounts of rain (> 100 mm) that fall during a short time period (< 6 hours) can cause flash floods, especially in places where the terrain supports accumulation of water as on low slope coast below a mountain. The ability of the operational ALADIN model to predict such cases is exemplarily illustrated by two cases.

On September 25, 2010, just after midnight an intensive rain hit Pula city on the southern part of the Istria Peninsula, Croatia. The rain was intensive for several hours and the rainfall rate measured by the ombrograph reached 43.9 mm per hour. Several rain-gauges in the area measured more than 150 mm/24h. The operational ALADIN forecast severely underestimated (the forecast maximum was below 10mm) the rainfall over the Istria peninsula during the night from September 24 to 25, 2010 (Fig 17a). The wind field is shown for 00 UTC on September 25, 2010 the time that is close to the period of the maximum rain intensity. The parallel suite rainfall structures were slightly better (Fig 17b) than the operational one, with second maximum of rainfall over the Istria peninsula reaching more than 35mm, but the predicted rainfall amount was far below the measured one. It is assumed that the observed severe precipitation was caused by convective activity supported by the synoptic conditions and/or local conditions that were not represented correctly in the initial conditions (as suggested by the result of the data assimilation forecast, Fig 17b) or the model was not able to represent its development. The last hypothesis has been tested using the non-hydrostatic set-ups of the ALADIN model and in higher resolution using different initial and boundary conditions (from 8 km runs with or without 3Dvar and using ARPEGE and IFS initial and boundary conditions) One run generated additional intensive rain band over an island 25 km east of Pula.

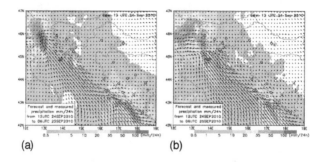

(a)   (b)

Figure 17. Accumulated precipitation from the operational (a) and data assimilation (b) forecast for 25th Sep 2010.

The flash flood hit Dubrovnik, Croatia, on November 22, 2010. The rain gauge measurements exceeded 100 mm/24h in the area and the one in Dubrovnik measured 161.4 mm/24h (the highest on record by that time), with a peak intensity of 71.5 mm/h (Fig 18b). The flood water level reached 1.5 m caused damage and endangered lives.

The operational and parallel suite model results forecast large 24 hour accumulated rainfall amounts in the area around Dubrovnik, with the maxima over 100 mm located above the surrounding areas of Montenegro and Bosnia and Herzegovina. The results of the high-resolution (2 km) non-hydrostatic ALADIN model run showed a dependency of the forecasts on the input data from initial and lateral boundary conditions. The position and time of the precipitation maxima in the high resolution ALADIN output fields (Fig 18a) was similar to the lower resolution (8 km) run used for initial and lateral boundary conditions. The peak intensity of precipitation in all model runs was late (3 hours in runs coupled to Arpege, 6 hours for IFS). The reason for this delay is yet unknown and could be attributed to the absence of measurements over the southern Adriatic.

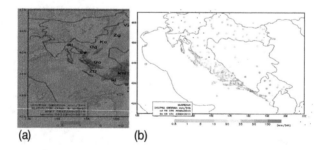

(a)   (b)

Figure 18. Accumulated 24 h precipitation from the 2 km resolution run (a) and measured (b) for 23rd Nov 2010.

3.6. Fog

Although far less dramatic than the previous ones, fog is also a weather event that can have a large impact since it is a nuisance to the traffic. The results of the operational model forecasts were not satisfactory for 2m temperature and cloudiness in a case of fog and low stratus accompanied with a temperature inversion in December 2004. The problems remained through several weeks of December 2004, as fog and low stratus remained covering the inland parts of Croatia, as well as Hungary and other valleys surrounding eastern Alps.

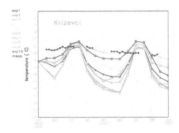

Figure 19. Comparison of the model 2m temperature evolution with the data measured (violet) in Križevci meteorological station. Exp1 to exp5 are obtained with the operational radiation scheme, and in exp6 to exp10 a new, enhanced, radiation scheme has been used. Experiments with random maximum overlap are shown with circles and those with random overlap are shown with squares. Open marks show experiments with the operational critical relative humidity profile and full marks show the experiments with the new critical relative humidity profile.

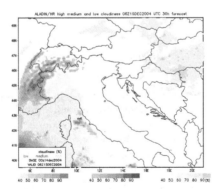

Figure 20. The operational forecast of low (shades of red), medium (shades of violet) and high (shades of blue) cloudiness for 06 UTC on December 15, 2004.

Subsequent operational model forecasts kept underestimating the cloud coverage with 30-60% on average when compared to observations by the crew from the synoptic stations. Consequently, the amplitude of the 2 m temperature diurnal cycle was much larger (5-10°C) than

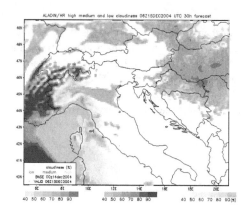

Figure 21. As Fig 20, but obtained from a simulation using the Xu-Randall cloudiness (Xu & Randall, 1996) scheme.

measured (Fig 19). Actually, the measured 2m temperature changed only slightly during the day (0-2°C), as well as from one day to another. This behaviour has made the 2m temperature forecast as useless as the cloudiness forecast.

These results have encouraged an extensive case study to find a model set-up that would produce an acceptable forecast of the low clouds as well as 2m temperature diurnal cycle (Tudor, 2010). The model initially recognizes the existence of the temperature inversion and the layer of air adjacent to the surface is almost saturated. But the cloud scheme diagnoses little cloudiness or fog, far less than exists in reality (Fig 20). Consequently, the radiation scheme heats the ground and breaks the inversion, making the situation even worse. In reality, a fog or low stratus layer keeps the longwave radiation from ground and reflects the shortwave radiation keeping the temperature of the surface layer almost constant. Longwave radiation flux divergence and shortwave reflection at the top of the cloud supports cooling and forma-tion of a temperature inversion.

The study (Tudor 2010) has shown that the combination of Xu-Randall cloud diagnosing scheme (Xu & Randall, 1996), new critical relative humidity profile, maximum overlap and operational (Ritter and Geleyn, 1982) radiation scheme produces forecast of low clouds and 2 m temperature of 1-3°C that is closest to the observed values. This result stayed nearly the same even when the more sophisticated radiation schemes were used. The introduction of prognostic microphysics improved the results too by increasing cloud coverage by 5-10% and reducing the temperature amplitude only slightly (up to 0.3°C). The introduction of more physical semi-Lagrangian horizontal diffusion (SLHD) has improved the fog and low stratus forecast close to terrain slopes by increasing cloud coverage by up to 50%. The numerical diffusion mixes the temperature and moisture along model levels that follow orography. Consequently, in the model the cloud layer from the valley is mixed with the clear air on the mountain ridge and the cloud in the valley reduces. Replacement of numerical diffusion with SLHD reduces this mixing and supports cloud formation on slopes of the valleys.

4. Conclusion

This chapter gives an overview on weather forecasting using the set-up of the NWP model ALADIN that is used for operational weather forecast in CMHS as an example for operation weather forecasting. ALADIN is a state-of-the-art modern NWP model. Using ALADIN we exemplarily discuss short-comings and challenges in modern operational weather forecasting. A high-resolution LAM is intended to predict the sub-synoptic weather features forced by topography or other local characteristics that can be absent in the main synoptic pattern. Successful prediction of these small-scale features enables usage of the LAM forecast in predicting the conditions important for the flight safety, vehicle road safety or navigation at sea. The operational suite has to be tuned in order to predict the high-impact weather events of local character that could be missing in the large scale forecast. The domain properties as well as the forecast model complexity are formed according to the needs of the forecast users and the computing capabilities.

High impact weather events can be of local origin, or determined by small scale characteristics particular to a certain region. Croatia is not spared from severe weather events, such as severe windstorms (bura and jugo wind), torrential rain and flash floods. Such events are often forecast in terms of probability via the ensemble prediction system of a global model. The severity of the event can be controlled by mechanisms not resolved by the global model, and dynamical adaptation of the individual EPS members improves the predictability of the severe weather events (Branković et al., 2007). The NWP section of CMHS has invested most effort in the deterministic high-resolution forecast based on analysis of numerous high impact weather cases.

The operational 8 km resolution forecast is run using DFI with the initial file from the global model ARPEGE since most of the measured data is assimilated in the global model. The data-assimilation suite has been established and runs parallel to the operational suite forecast. This parallel run has revealed the benefits of doing the data-assimilation in higher resolution (8 km) than in the global model (see the Pula flash flood case).

Forecasting the formation and distribution of fog is controlled by subtle variations in humidity and temperature fields and require maintenance of a sensitive balance between the advection, radiation and turbulent fluxes. The profound influence of topography on the atmospheric flow reflects itself in a number of features that affect local weather, such as upslope, downslope and gap winds, coastal barrier jets as well as land sea breezes.

The operational high-resolution (2 km) forecast of the 10 m wind is achieved through the dynamical adaptation method adapted from Žagar & Rakovec (1999). The prediction of severe wind variability and strength is improved in weather situations with strong flow over complex topography, like bura in Croatia (Ivatek-Šahdan & Tudor, 2004). This high resolution wind forecast has provided warnings in numerous cases severe wind as well as to forecast the road conditions, assess the wind energy potential (Horvath et al., 2011) and has been used in numerous applications.

The forecast abilities of HRDA and similar packages are limited to the severe windstorms related to a synoptic forcing. Omitting moist and radiation processes prevents it from being useful in forecasting other phenomena could benefit from the high resolution, such as local convective storms. The windstorm that is a consequence of a local pressure disturbance requires a full forecast run in 2 km resolution using more complex NH ALADIN model set-up (Tudor & Ivatek-Šahdan, 2010). Several air crash investigations have revealed that this "full-run" high resolution ALADIN forecast enables the prediction of lee waves and zones of increased turbulence as well as icing zones.

These flaws of the operational suite have encouraged the introduction of the 2 km resolution 24 hour forecast with NH ALADIN set-up using the complete set of physics parameterizations to the operational suite at the beginning of summer 2011. Several case studies of high-impact weather phenomena have been used to set-up the model configuration used operationally. The results from these studies have provided encouraging results.

Author details

Martina Tudor, Stjepan Ivatek-Šahdan, Antiono Stanešić, Kristian Horvath and Alica Bajić

Croatian Meteorological and Hydrological Service, Croatia

References

[1] ALADIN International Team, (1997) The ALADIN project: Mesoscale modelling seen as a basic tool for weather forecasting and atmospheric research. *WMO Bull.*, 46, 317–324.

[2] Bajić, A. (2003) Očekivani režim strujanja vjetra na autocesti Sv. Rok (jug) – Maslenica. *Građevinar*, 55, 149-158.

[3] Bajić, A., Ivatek-Šahdan, S. & Horvath, K. (2007) Spatial distribution of wind speed in Croatia obtained using the ALADIN model. *Cro. Met. J.* 42 , 67–77.

[4] Bajić, A., Ivatek-Šahdan, S., Žibrat, Z. (2008) ANEMO-ALARM iskustva operativne primjene prognoze smjera i brzine vjetra. *GIU Hrvatski cestar.* 109-114

[5] Berre L. (2000) Estimation of Synoptic and Mesoscale Forecast Error Covariances in a Limited-Area Model, *Mon. Wea. Rev.* 128, 644-667.

[6] Bölöni, G., Horvath K. (2010) Diagnosis and tuning of the background error statistics in a variational data assimilation system. *Időjárás* 114, 1-19.

[7] Branković, Č., Matjačić, B., Ivatek-Šahdan, S. & Buizza, R. (2007) Dynamical downscaling of ECMWF EPS forecasts applied to cases of severe weather in Croatia. *ECMWF RD Technical Memorandum 507*, 38 pp.

[8]   Branković, Č., Matjačić, B., Ivatek-Šahdan, S. & Buizza, R. (2008) Downscaling of ECMWF Ensemble Forecasts for Cases of Sevére Weather: Ensemble Statistics and Cluster Analysis. *Mon.Wea. Rev.* 136, 3323–3342.

[9]   Brozkova R, Klarić, D., Ivatek-Šahdan, S., Geleyn, J.F., Casse, V., Siroka, M., Radnoti, G., Janousek, M., Stadlbacher, K., Seidl, H. (2001) DFI blending: an alternative tool for preparation of the initial conditions for LAM. WGNE *Blue Book.* 31, 1.7-1.8.

[10] Catry B., Geleyn J.-F., Tudor M., Bénard P. & Trojakova A. (2007). Flux conservative thermodynamic equations in a mass-weighted framework. *Tellus* 59A, pp 71–79

[11] Courtier, 1999: Data assimilation concepts and methods, *ECMWF lecture notes, European Centre for Medium-Range Weather Forecasts*, Reading, England, 59 pp.

[12] Davies, H. C. (1976) A lateral boundary formulation for multilevel prediction models. *Quart. J. Roy. Meteor. Soc.*, 102, 405–418.

[13] Davies, H. (1983) Limitations of some common lateral boundary schemes used in regional NWP models. *Mon. Wea. Rev.*, 111, 1002–1012.

[14] Durran, D.R. (1999) Numerical methods for wave equations in geophysical fluid dynamics. *Springer*, pp 465.

[15] Geleyn J.-F. (1988). Interpolation of wind, temperature and humidity values from model levels to the height of measurement. *Tellus*, 40A, pp.347–351

[16] Geleyn, J.-F., Bazile, E., Bougeault, P., Déqué, M., Ivanovici, V., Joly, A., Labbé, L., Piédélièvre, J.-P., Piriou, J.-M., Royer, J.-F. (1994) Atmospheric parametrizations schemes in Meteo-France's ARPEGE NWP model. *ECMWF seminar proceedings on Parametrization of sub-grid scale physical processes*, pp. 385-402.

[17] Geleyn J.-F. & Hollingsworth A (1979) An economical analytical method for the computation of the interaction between scattering and line absorption of radiation. *Beitr Phys Atmos* 52:1–16.

[18] Geleyn J.-F., Benard P. & Fournier, R. (2005a) A general-purpose extension of the Malkmus band-model average equivalent width to the case of the Voigt line profile. *Quart. J. Roy. Meteor. Soc.* 131:2757–2768

[19] Geleyn J.-F., Fournier R., Hello G., Pristov N. (2005b) A new 'bracketing' technique for a flexible and economical computation of thermal radiative fluxes, scattering effects included, on the basis the Net Exchanged Rate (NER) formalism. WGNE Blue Book

[20] Geleyn J.-F., Vana F., Cedilnik J., Tudor M. & Catry B. (2006). An intermediate solution between diagnostic exchange coefficients and prognostic TKE methods for vertical turbulent transport. *WGNE Blue Book*

[21] Geleyn J.-F., Catry B., Bouteloup Y. & Brožkova, R. (2008). A statistical apreach for sedimentation inside a microphysical precipitation scheme. *Tellus* 60A,649–662

[22] Gerard, L. (2007) An integrated package for subgrid convection, clouds and precipitation compatible with the meso-gamma scales. *Quart. J. Roy. Meteor. Soc.*, 133, 711–730.

[23] Gerard, L. & Geleyn, J.-F. (2005) Evolution of a subgrid deep convection parametrization in a limited area model with increasing resolution. *Quart. J. Roy. Meteor. Soc.*, 131, 2293–2312.

[24] Gerard, L., Piriou, J.-M., Brožková, R., Geleyn, J.-F. & Banciu, D. (2009) Cloud and Precipitation Parameterization in a Meso-Gamma-Scale Operational Weather Prediction Model. *Mon. Wea. Rev.*, 137, 3960–3977.

[25] Giard, D. and Bazile, E. (2000) Implementation of a new assimilation scheme for soil and surface variables in a global NWP model. *Mon. Wea. Rev.* 128, 997-1015.

[26] Gospodinov, I., Spiridonov, V., & Geleyn, J.-F., (2001) Second order accuracy of two-time-level semi-Lagrangian schemes. *Quart. J. Roy. Meteor. Soc.*, 127, 1017–1033.

[27] Grisogono, B., Belušić, D. (2009) A review of recent advances in understanding the meso- and microscale properties of the severe Bora wind. *Tellus* 61A, 1–16.

[28] Grubišić, V. (2004) Bora-driven potential vorticity banners over the Adriatic. *Quart. J. Roy. Meteor. Soc. 130*, 2571-2603.

[29] Haugen, J. E. & Machenhauer, B. (1993) A spectral limited-area model formulation with time-dependent boundary conditions applied to the shallow-water equations. *Mon. Wea. Rev.*, 121, 2618–2630.

[30] Hdidou, F.Z., (2006) Test of an assimilation suite for ALBACHIR based on the variational technique. available at www.wmo.int

[31] Hólm, E.V., 2008: Lecture notes on assimilation algorithms. *ECMWF, European Centre for Medium-Range Weather Forecasts*, Reading, England, 30 pp.

[32] Hollingsworth, F. Rabier and M. Fisher, 1998: The ECMWF implementation of three-dimensional variational assimilation (3d-Var). I: Formulation. *Quart. J. Roy. Meteor. Soc.*, 124, 1783-1807.

[33] Horvath, K., Bajić, A., & Ivatek-Šahdan, S. (2011) Dynamical Downscaling of Wind Speed in Complex Terrain Prone To Bora-Type Flows. *J. Appl. Meteor. Climatol.*, 50, 1676–1691.

[34] Horvath, K., Ivatek-Šahdan, S., Ivančan-Picek, B. & Grubišić, V. (2009) Evolution and structure of two severe cyclonic bora events: contrast between the northern and southern Adriatic. *Weather and forecasting* 24, 946-964.

[35] Ivatek-Šahdan, S. & Ivančan-Picek, B. (2006) Effects of different initial and boundary conditions in ALADIN/HR simulations during MAP IOPs. Meteorol. Z. 15, 187–197.

[36] Ivatek-Šahdan, S. & Tudor M. (2004) Use of high-resolution dynamical adaptation in operational suite and research impact studies. *Meteorol Z* 13(2):1–10

[37] Kann A., Seidl H., Wittmann C. & Haiden T. (2009) Advances in predicting continental low stratus with a regional NWP model. *Wea. Forecasting*, 25, 290–302.

[38] Kessler E. (1969) On distribution and continuity of water substance in atmospheric circulations. *Meteorol Monogr Am Meteorol Soc* 10(32):84

[39] Louis J.-F., Tiedke M. & Geleyn J.-F. (1982) A short history of PBL parameterization at ECMWF. In: *Proceedings from ECMWF workshop on planetary boundary layer parameterization, 25–27 November 1981*, pp 59–79

[40] Lynch, P., X.-Y. Huang (1994) Diabatic Initialization using recursive filters. – *Tellus* 46A, 583–597.

[41] Machenhauer, B., J.E. Haugen (1987) Test of a spectral limited area shallow water model with timedependent lateral boundary conditions and combined normal mode/ semi-lagrangian time integration schemes. *Techniques for Horizontal Discretization in Numerical Weather Prediction Models, 2–4 November 1987*, ECMWF, 361–377.

[42] Mahfouf, J.,F., 1991: Analysis of soil moisture from near-surface variables: A feasibility study. *J. Appl. Meteor.*, 30, 1534-1547

[43] Mlawer E.J., Taubman S.J., Brown P.D., Iacono M.J. & Clough S.A. (1997) Radiative transfer for inhomogeneous atmospheres: RRTM, a validated correlated-k model for the longwave. *J Geophys Res* 102D:16663–16682

[44] Morcrette J.-J. (1989) Description of the radiation scheme in the ECMWF Model. *Tech Memo 165, ECMWF*, 26 pp

[45] Noilhan, J., Planton, S. (1989) A simple parapetrization of land surface processes for meteorological models. *Mon. Wea. Rev.* 117, 536-549.

[46] Pauluis O., Emanuel K. (2004) Numerical instability resulting from infrequent calculation of radiative heating. Mon Weather Rev 132:673–686

[47] Pielke, R.A. (2002) Mesoscale meteorological modelling. *Academic press*, pp 676.

[48] Radnóti, G. (1995) Comments on "A spectral limited-area formulation with time-dependent boundary conditions applied to the shallow-water equations". *Mon. Wea. Rev.*, 123, 3122– 3123.

[49] Redelsperger J.L., Mahé F., Carlotti P. (2001). A simple and general subgrid model suitable both for surface layer and free-stream turbulence. *Bound.-Layer Meteor.* 101, pp.375–408.

[50] Ritter B. & Geleyn J.-F. (1992) A comprehensive radiation scheme for numerical weather prediction models with potential applications in climate simulations. *Mon Wea. Rev* 120:303–325

[51] Robert, A. (1982) A semi-Lagrangian and semi-implicit numerical integration scheme for the primitive equations. *J. Meteor. Soc. Japan*, 60, 319-325.

[52] Simmons, A.J., Burridge, D.M. 1981: An Energy and Angular-Momentum Conserving Vertical Finite-Difference Scheme and Hybrid Vertical Coordinates. *Mon. Wea. Rev.* 109, 758-766.

[53] Staniforth A, Côté J. (1991) Semi-Lagrangian integration schemes for atmospheric models – A review. *Mon. Wea. Rev.* 119 2206-2223.

[54] Termonia, P. (2008) Scale-selective digital filtering initialization. *Mon. Wea. Rev.*, 136, 5246-5255.

[55] Tudor M. (2010). Impact of horizontal diffusion, radiation and cloudiness parameterization schemes on fog forecasting in valleys. *Met. Atm. Phy.* Vol.108, pp. 57-70.

[56] Tudor, M. (2011). The meteorological aspects of the DART field experiment and preliminary results. *Cro. Met. J.* 44/45, 31-46.

[57] Tudor, M. & Ivatek-Šahdan, S. (2002) The MAP-IOP 15 case study. *Cro. Met. J.* 37, 1-14.

[58] Tudor, M. & Ivatek-Šahdan, S. (2010) The case study of bura of 1st and 3rd February 2007, *Meteorol. Z.*, 19, pp. 453-466.

[59] Tudor, M., Termonia, P., (2010) Alternative formulations for incorporating lateral boundary data into limited-area models. *Mon. Wea. Rev.* 138, pp. 2867-2882.

[60] Unden, P., Rontu, L., Jarvinen, H., Lynch, P., Calvo, J., Cats, G., Cuxart, J., Eerola, K., Fortelius, C., Garcia-Moya, J.A., Jones, C., Lenderlink, G., McDonald, A., McGrath, R., Navascues, B., Nielsen, N., Odegaard, V., Rodriguez, E., Rummukainen, M., Room, R., Sattler, K., Hansen Sass, B., Savijarvi, H., Schreur, B., Sigg, R., The, H., Tijm, S., (2002) HIRLAM-5 scientific documentation. *SMHI, S-601 76 Norrkoping, Sweden*, pp 146, available at www.hirlam.org

[61] Váňa F., Bénard P., Geleyn J.-F., Simon A. & Seity Y. (2008). Semi-Lagrangian advection scheme with controlled damping–an alternative way to nonlinear horizontal diffusion in a numerical weather prediction model. *Quart. J. Roy. Meteor. Soc*, Vol.134, pp. 523-537.

[62] Wang, Y., Bellus, M., Wittmann, C., Steinheimer, M., Weidle, F., Kann, A., Ivatek-Šahdan, S., Tian, W., Ma, X., Tascu, S., Bazile, E. (2011) The Central European limited-area ensemble forecasting system: ALADIN-LAEF. *Quart. J. Roy. Meteor. Soc* 137, 483-502.

[63] Warner, T., Peterson, R., Treadon, R. (1997) Atutorial on lateral boundary conditions as a basic and potentially serious limitation to regional numerical weather prediction. *Bull. Amer. Meteor. Soc.*, 78, 2599-2617.

[64] Xu K.-M. & Randall D.A. (1996) A semi-empirical cloudiness parameterization for use in climate models. *J Atmos Sci* 53:3084–3102

[65] Yessad, K (2011) Library architecture and history of the technical aspects in AR-PEGE/IFS, ALADIN and AROME in the cycle 37 of ARPEGE/IFS. *Meteo-France*, pp. 22, available at http://www.cnrm.meteo.fr/gmapdoc/IMG/pdf/ykarchi37t1.pdf

[66] Zaninović, K., Gajić-Čapka, M., Perčec Tadić, M., Vučetić, M., Milković, J., Bajić, A., Cindrić, K., Cvitan, L., Katušin, Z., Kaučić, D., Likso, T., Lončar, E., Lončar, Ž., Mihajlović, D., Pandžić, K., Patarčić, M., Srnec, L., Vučetić, V., (2008) Climate atlas of Croatia: 1961. - 1990. : 1971. - 2000. . Zagreb : CMHS *Monograph*, pp. 200.

[67] Žagar, M., Rakovec, J. (1999): Small-scale surface wind prediciton using dynamical adaptation. *Tellus*, Vol.51A, pp. 489–504.

Climate Modeling

Chapter 4

Grids in Numerical Weather and Climate Models

Sarah N Collins, Robert S James, Pallav Ray,
Katherine Chen, Angie Lassman and
James Brownlee

Additional information is available at the end of the chapter

1. Introduction

Since the early 20th century numerical weather prediction (NWP) has increasingly become one of the most important and complicated problems of modern science. With the advent of computers, increased observations, and progress in theoretical understanding, numerical models were developed. Since then, such models are playing an increasing role in understanding and predicting weather and climate and have been a driving force in the advancement of the meteorological sciences. Numerical models are a mathematical representation of the earth's climate system including the atmosphere, ocean, cryosphere and land, among others. The models divide the area of interest into a set of grids and then make use of observations of variables such as surface pressure, winds, temperature and humidity at numerous locations throughout the globe. The observed values are then assimilated and used by the model to predict future evolution of the earth's weather and climate. In the mid 20th century, models evolved from a simple model with a single atmospheric layer to a multi-layer primitive equation model capable of predicting cyclone development [1].

Due to the amount of computer processor time, memory, and disk storage required to run numerical models, the atmosphere cannot be represented perfectly by the model and thereby is approximated by a finite data set. The atmosphere is represented in a model by a three-dimensional set of points, called grids that cover the region of interest. Figure 1 demonstrates the importance of the number of discrete grid points in order for the model to best represent the atmospheric structures. Figure 1a uses a grid spacing of 1 in. whereas Figure 1b uses a grid spacing of 0.5 in. Comparison of the two figures shows how increasing the number of grid points allows the model to better represent the actual wave function. As the number of discrete grid points increases, so increases the representation of the atmosphere. The NWP models in the 1950s had grid points every few hundred kilometers in the horizontal, whereas today, models used in operational forecasting have grid points every 10-100 km. The horizontal distance between the adjacent grids are often known as grid-spacing or resolution. Many models have the ability to nest finer grids within a coarse grid resulting in a nested grid with much higher resolution. The vertical resolution of numerical models improved alongside horizontal resolution, such that models today can have more than 50 vertical layers.

The resolution of a model also depends on the area coverage. General circulation models (GCMs), which are global, typically have coarse resolutions and are necessary for long-range forecasts. Regional models often called limited area models, have finer resolution and are used for short-range forecasts. Limited area models can be run closer to real time because they can be initialized using local observations, but still depend on other model output for boundary conditions. On the other hand, global models must wait for observations from around the globe for their initializations. With the increase in computer power, high-resolution regional models are covering larger area, whereas, global models are having finer resolutions. It is expected that cloud-system-resolving global models will be common in the coming decades.

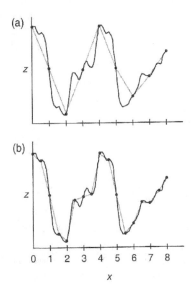

Figure 1. Idealized grid point approximation (gray) of a function z (black) plotted on the interval [0, 8]. Grid points every 1 in (a) and every 0.5 in. (b) are indicated by black dots. (Source: [1])

An integral part of any weather and climate model is the numerical schemes that are designed to convert the set of partial differential equations (PDE's) that represent geophysical equations into a set of algebraic expressions. This is essential because computers cannot solve PDE's directly; but can perform algebraic operations like addition and multiplication. The conversion of PDE's to simple algebraic equations involves primarily two steps; computation of a derivative and representation of the solution by a finite data set. This can be achieved using two different methods:

i.   Series expansion method: In this method, an unknown function is approximated by a linear combination of a finite set of continuous expansion functions. When the expansion functions are orthogonal, the method is known as the spectral method. When the expansion functions are non-zero in only a small part of the local domain, the method is called the finite element method. The finite element method is generally not as efficient as the spectral method because it usually requires the solution of implicit algebraic equations and is mostly time-independent. However, there are certain situations when the finite element method is useful, such as approximating the vertical structure of the atmosphere or in steady state situations [2]. The spectral method is used extensively for global medium-range forecasting models and global climate models. Gates [3] noted that out of 29 models that participated in AMIP (Atmospheric Model Inter-comparison Project), 19 were spectral.

ii.   Grid-point method: In this method, the equations are approximated on grid points using a finite set of data. Finite difference is an example of the grid-point method. Series expansion and grid-point methods are used extensively for spatial derivatives [4]. However, the time derivative is almost always approximated using the finite difference method. Most short-range and regional models are grid-point models.

The development of modern discretization techniques like that of semi-implicit and the semi-Lagrangian schemes, have kept the cost of numerical modeling in check by having less stringent stability conditions on the time-step and more accurate space discretization [4, 5].

The chapter is organized as follows: Section 2 documents a brief history of the model grids. Section 3 describes different shapes of grids, followed by staggering of grids in section 4. Further discussions are presented in section 5, followed by conclusions in section 6.

2. A brief history of model grids

Earlier days of real time numerical weather prediction were far from what is now considered "the greatest intellectual achievement and scientific advancement in twentieth century atmospheric science" [4]. Bjerknes [6] recognized for the first time that NWP is an initial value problem. Prior to Bjerknes, however, Cleveland Abbe, who was the first chief meteorologist of the US weather bureau, prepared and issued the first official weather forecast in the US on February 19, 1871. He introduced standard forecast verification procedures and wrote a paper on "The Physical basis of Long-range Weather Forecasts" [7], where he correctly wrote the set of primitive equations that govern the atmospheric motions.

The idea of forecasting weather by numerical process using physically based models was first proposed by Lewis Fry Richardson [8]. Following the basic idea of an initial value problem, if the values of certain environmental variables are known, then the physical equations can be used to calculate their values at a time in the future. Richardson was able to simplify the equations of motions presented by Abbe and later by Bjerknes using his knowledge of meteorology and mathematics. Richardson proposed to divide the earth's surface into a grid, with each grid cell the base of a vertical column of the atmosphere. Each vertical column was then divided into several layers, resulting in a three-dimensional grid of atmospheric boxes. In Figure 2, each column (or grid box) extends 3° in the east-west direction and 200 km in the north-south direction. This resulted in 12,000 columns to cover the entire globe. To test his technique and provide an example of how to use it, Richardson performed the calculations for two adjacent columns [8]. Each column was divided into five layers at heights of 2, 4.2, 7.2, and 11 km (or about 800, 600, 400 and 200 hPa) above sea-level and the values of the variables were fixed at the center of each box (Figure 3). He computed only the initial tendency at a single point for pressure at the base of each layer, temperature at the stratosphere, water content at the lower four layers and the two components of horizontal wind. His calculation of change of pressure at the point considered was 145 hPa in 6 hrs., an obviously unrealistic value. His forecast failed as a result of short-period oscillations called gravity waves that created "noise" in the observed data set, thereby causing error in the initial conditions used in his forecast.

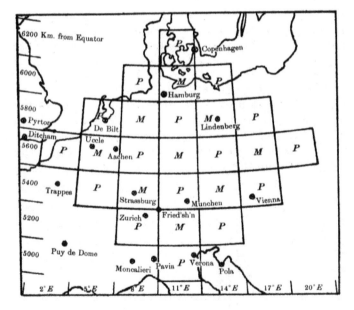

Figure 2. Grid used by Richardson in 1922 to calculate the pressure change in central Germany. X-axis shows the longitude and Y-axis shows the distance (in km) from the equator. Each grid box is 3° in the longitudinal direction and 200 km in the latitudinal direction. (Source: [9])

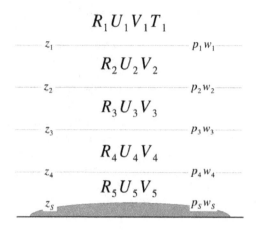

$$R_1 U_1 V_1 T_1$$

z_1 ──────────────── $p_1 w_1$

$$R_2 U_2 V_2$$

z_2 ──────────────── $p_2 w_2$

$$R_3 U_3 V_3$$

z_3 ──────────────── $p_3 w_3$

$$R_4 U_4 V_4$$

z_4 ──────────────── $p_4 w_4$

$$R_5 U_5 V_5$$

z_S ──────────────── $p_S w_S$

Figure 3. Vertical grid used by Richardson (1922).

3. Shapes of grids

Richardson's effort of predicting weather using grid points (as seen in Figures 2, 3) set the stage for future development of grids in different shapes. In order to accommodate the spherical shape of the earth and represent the equations more accurately and efficiently, there are different grid shapes used in numerical models; e.g., rectangular, triangular, and hexagonal (Figure 4). A brief description of these grids along with their relative advantages and disadvantages are presented next with an emphasis on the following criteria: (i) Suitability for cloud-scale to global-scale; (ii) Efficiency on different computer architectures and scalability on massively parallel computers; (iii) Conservation of mass and other quantities; and (iv) Capability of local grid refinement and regional domains. However, the method of distribution of grid points over the sphere is yet to be solved in a fully satisfactory manner.

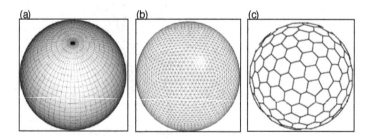

Figure 4. Examples of (a) rectangular or latitude-longitude grid, (b) triangular grid and (c) hexagonal grid.

3.1. Rectangular/square grids

The rectangular/square (or latitude-longitude) grid is the most commonly used grid in the NWP models (e.g., [10]). The rectangular grid is simple in nature but suffers from "the polar problem" where the lines of equal longitude, known as meridians, converge to points at the poles (Figure 5a). The poles are unique points and may cause violation of global conservation laws within the model. To maintain computational stability near the poles, small integration time-steps could be used, but at great expense. The high resolution in the east-west direction near the poles would be wasted because the model uses lower resolution elsewhere ([11]).

(a)   (b)

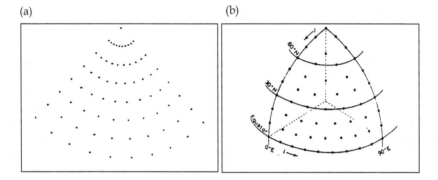

Figure 5. (a) Demonstration of the polar problem in which the meridians converge to a single point at the poles and (b) its remedy using the Kurihara grid.

A rotated grid can overcome the polar problem for limited area models ([4]), whereas for global models, other grid shapes are used. For example, Kurihara [12] proposed to use 'skipped' or 'Kurihara' grid (Figure 5b). Unfortunately use of the Kurihara grid causes spuriously high pressure to develop at the poles. As a result, their use has been severely limited or abandoned in finite difference models. However, problems due to the use of the Kurihara grid can be resolved by using more accurate numerical schemes [13]. In the late 60's and early 70's, the application of quasi-uniform grids was proposed as a method to avoid the polar problem of the grid-point models [14]. For example, the Global Forecasting System (GFS) model has roughly a square grid near the equator, a more rectangular grid in the mid-latitudes, and a triangular grid near the poles, eventually converging to a point at the poles. Another example of a model that uses the rectangular grid type is the North American Mesoscale Model (NAM). When compared to the resolution of the GFS, the NAM does not have a grid stretching problem since the model calculates variables close to the poles. This is due to the NAM not relying on a latitude-longitude system for creating its grid bounds and deferring to a more precise horizontal measurement system. The other problem with the latitude-longitude grid is the need for special filters to deal with the pole singularities. They also do not scale well on massively parallel computers.

3.2. Triangular grids

Triangular grids are not used as often in models as are rectangular grids. One form of quasi-uniform grid whose base element is a triangle is the spherical geodesic grid. Icosahedral grids, first introduced in the 1960s, give almost homogeneous and quasi-isotropic coverage of the sphere. The grid is made by dividing the triangular faces of an icosahedron into smaller triangles, the vertices of which are the grid points. Each point on the face or edge of one of the faces of the icosahedron is surrounded by six triangles making each point the center of a hexagon. The triangular faces of the icosahedrons are arranged into pairs to form rhombuses, five around the South Pole and five around the North Pole. The poles are chosen as two pentagonal points where the five rhombuses meet. The main advantage of the geodesic grid is that all the grid cells are nearly the same size. The uniform cell size allows for computational stability even with finite volume schemes.

3.3. Hexagonal grids

Similar to triangular grids, hexagonal grids are also not used as often as the rectangular/square grids. In this method, variables are calculated at each grid intersection between different hexagons, in addition to being calculated in the center of the hexagonal grid. Sadourney et al. [15] describes in detail how the spherical icosahedral-hexagonal grid is constructed. They solved the non-divergent barotropic vorticity equation with finite difference methods on the icosahedral-hexagonal grids. Majewski et al. [16] utilizes an approach that uses local basis functions that are orthogonal and conform perfectly to the spherical surface. A study done by Thuburn [17] shows a method of creating a global hexagonal grid, but then using a finite differencing method to calculate the rate-of-change of different variables without having to create triangles within the hexagonal grids. The space differencing scheme using the icosahedral-hexagonal grid gives a satisfactory approximation to the analytical equations given an initial condition and remains nonlinearly stable, for any condition.

Thuburn [17] also noted that his method may not be as accurate as those which included an additional point in the center of the hexagonal grid, but his method was computationally faster, and was able to accurately depict the polar regions since there was no need of stretching the grid in that region. The other advantages of the hexagonal grid are: (i) Removes the polar problem. (ii) Permits larger explicit time steps. (iii) Most isotropic compared to other grid types. (iv) Conservation of quantities in finite volume formulation. (v) Can be generalized easily to arbitrary grid structures.

4. Grid staggering

After the choice of distribution of grid points (i.e., rectangular, triangular, hexagonal, etc.), the next step is to arrange the prognostic variables on the grid. When all the prognostic variables are defined at the same point in a grid, it is called an unstaggered grid. On the other hand, when prognostic variables are defined at more than one point in a grid, it is called a staggered grid. Characteristically a staggered grid has the values of the wind components at separate points than the thermodynamic variables within the grid cell. Consequently, a model's resolution is defined as the average distance between adjacent grid points with the same variables.

Staggering is not only performed in the horizontal direction, but also in the vertical direction, as well as in time, and any combinations thereof.

4.1. Horizontal staggering

There are a variety of different methods in which models calculate temperature, winds, surface pressure, geopotential, and other meteorological variables within their grids. Five different types of grids were introduced by Arakawa and Lamb [18] and are shown in Figure 6. Of these grids, A is an unstaggered grid where the variables are defined at the same points, e.g., at the center or at the corners of the grid. Grids B through E are all staggered grids where the variables are defined at different points. Since all variables are defined at all the grid points in the A grid, it is easy to construct a higher-order accurate scheme. The main disadvantage is that the differences are computed over a distance of $2\Delta x$, and the adjacent points are not coupled for the pressure and convergence terms. In grid B, evaluation of the two sets of variables are at different points, e.g., one might evaluate the velocities at the center of a grid and masses at the grid corners. Since the B and E grids have wind components at the same point, they are often called semi-staggered. In grid C, velocities are calculated at the mid-point between grid cells and h is calculated at the corners (or intersections of grid cells). The main advantage of the C grid is that the pressure and convergence terms are computed over a distance Δx, which is half of that in the A grid indicating a doubling of the resolution compared to the A grid. Most non-hydrostatic mesoscale models like fifth generation mesoscale model (MM5) and Weather Research and Forecasting (WRF) use the C grid. The B grid was chosen in the UK Met Office (UKMO) unified model [19, 20] for climate simulation as well as numerical weather prediction.

From the grids presented in Figure 6 it can be seen that the resolution of models does not depend solely on the size or shape of the grid but also the location in which the model calculates various atmospheric variables. As the number of grid cells increases, the differences in effective resolution between different grid types eventually goes to zero as the number of grid cells goes to infinity [21].

In Figure 6, grid D is a slight variation of grid C, with the u and v variables being oriented with a rotation of 90°. This variation allows for a simple evaluation of the geostrophic wind. This is done by creating better averages for variables such as pressure gradient, mass convergence/divergence, and the Coriolis terms [21]. The D grid was used (with time staggering) in National Meteorological Center's (NMC's) nested grid model [22]; however, this grid is no longer used in any popular atmospheric model because of no added benefit.

The staggered E grid is rotated 45° relative to the B grid, but has an increased grid-spacing (Figure 6e) compared to the B grid. One problem with this grid is when the domain is small and one-dimensional, this grid is equivalent to grid A, but with less computational efficiency. National Centers for Environmental Prediction (NCEP) eta model uses grid E [23].

Although staggered grids have higher equivalent resolution than unstaggered grids, they are also more complex. Overall, the C grid is becoming more popular in recent times with the E grid its closest competitor.

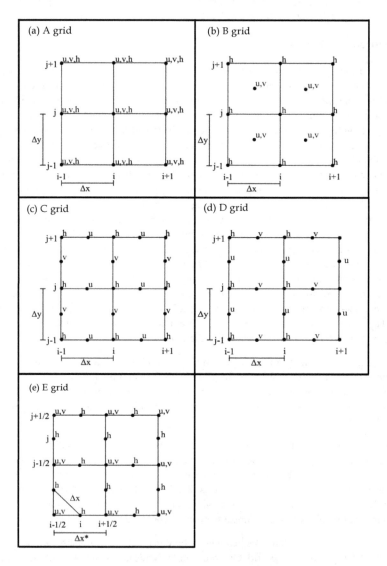

Figure 6. Grid types based on horizontal staggering, namely, (a) the A grid, (b) the B grid, (c) the C grid, (d) the D grid and (e) the E grid. Δx is the east-west resolution and Δy is the north-south resolution.

4.2. Vertical staggering

Staggering of grids in the vertical direction also provides certain advantages. For example, vertical staggering introduced by Lorenz [24] maintains the requirement of boundary conditions of no flux at the top and bottom (Figure 7a). However, the Lorenz grid allows the formation of a spurious computational mode [25]. This problem does not exist in the Charney-Phillips grid ([26]; Figure 7b) in which the vertical staggering, being more consistent (compared to Lorenz grid) with hydrostatic equation, does not allow the additional computational mode [27]. Most of the present state-of-the-art numerical models have staggered grids in the vertical direction with prognostic variables at the center of the layer and the vertical velocity at the boundary of the layers. Such an example is shown in Figure 8 from the WRF model [28].

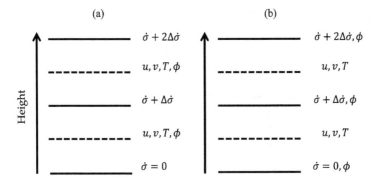

Figure 7. Staggered grids in the vertical following (a) Lorenz [24] & (b) Charney-Phillips [26].

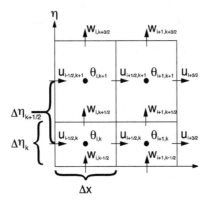

Figure 8. Vertically staggered grid from the WRF model. (from [28])

4.3. Time staggering

Staggering of grids is not confined in space, staggering can also be in time. For example, for atmospheric flow using the leapfrog scheme grid D is ideal when staggered in time (Figure 9). Time staggering was first introduced by Eliassen [29], which involves defining variables at every second time step on an offset D grid. A slight variation of this approach performed by Bratseth [30] used a higher-order interpolation to transfer values back from the offset grid to the original grid. All the differences are calculated on a distance Δx. Despite this advantage, such time staggering is not used due to the complexities that arise from the additional staggering and need of special procedures for starting the leapfrog scheme.

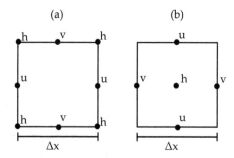

Figure 9. Staggered grids in space and time with (a) even time steps (b) odd time steps

5. Discussions

The ability of the NWP model to accurately represent atmospheric phenomena is based on three conditions; scientific knowledge, the availability of observational data, and computer processing abilities. If enough observational data is available and enough scientific knowledge is present then the limiting factor of an accurate forecast is the power of the processing computer. There are other issues of relevance that warrant further discussion.

5.1. Grid splitting

The B and E grids discussed in section 4.1 can be considered as being made up of two C grids. An example is shown in Figure 10 for the B grid. Ignoring the distinction between variables in upper- and lower-case characters, the figure represents a B grid. Considering only the lower-case characters, the figure represents a C grid whose axes are rotated 45° counterclockwise relative to that of the B grid. On the other hand, considering only the upper-case characters, the figure represents a second C grid that is shifted by one grid-length along the x-axis of the B grid. Precaution must be taken in formulating B- and E-grid models to avoid solutions splitting into two separate distributions on the two C grids [31].

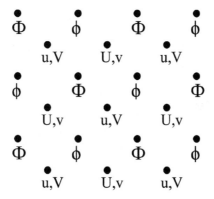

Figure 10. Representation of variables on the B grid, which can be seen as a combination of two C grids rotated through 45°, denoted by variables in upper and lower case.

5.2. Nested grids

Some models are run with finer-resolution grids nested inside coarser-resolution grids within the same model. Grid nesting is used when computational limitations prohibit fine-resolution grids from covering the entire model domain. Nesting can be one-way or two-way. Information in a two-way nested grid is shared both ways, from the coarse-grid to the fine-grid and from the fine-grid to the coarse-grid. In Figure 11, where the fine-grid covers the coarse-grid, the forecast variables for the coarse-grid are updated based on the fine-grid prediction. The coarse-grid prediction provides boundary conditions on the nest interface for use in the fine-grid prediction. Advantages of the two-way nested grid include, fine-scale processes resolved on the finer grid are allowed to affect the larger-scale flow on the coarse-grid. This is important for numerical weather prediction because the small-scale processes in the atmosphere greatly influence the large-scale processes in the atmosphere. Since the predictions on coarse-resolution grids take less computer time and memory compared to fine-resolution grids, the outermost boundary of the model can be moved far from the forecast region, while the fine-resolution domain remains small enough to run in real time. Moving nests are also common in the present models where a higher resolution nest can move with the phenomenon of interest (e.g., hurricane) to provide details that wouldn't be possible in a coarse resolution simulation.

An example of staggered C grid with nesting is also shown in Figure 12.

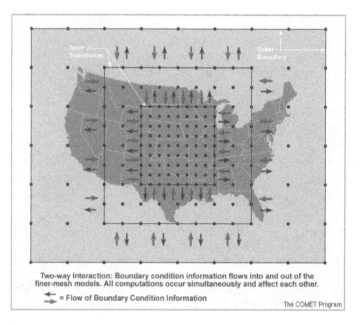

Figure 11. Example of a two way nested grid with coarse resolution outer domain and finer resolution inner domains. Staggering type is typically same for all domains. The arrows indicate direction of information exchange. (Source: *www.comet.ucar.edu)*

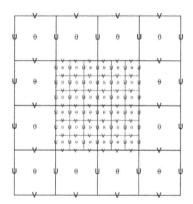

Figure 12. A portion of a nested grid with C grid staggering with 3:1 grid size ratio. The solid lines denote coarse-grid cell boundaries, and the dashed lines are the boundaries for each fine grid cell. The bold typeface variables along the interface between the coarse- and the fine-grid define the locations where the specified lateral boundaries for the nest are in effect. (Taken from [28]).

5.3. Mesh refinement

Adaptive mesh refinement is a method where model resolution is refined and the solution has fine-scale structure, rather than in a fixed region as done in a conventional nesting. This is being used in the model for prediction across scales [32, 33, 34]. It has all the advantages of a hexagonal grid as described in section 3.3. It uses centroidal Voronoi tessellations with a C grid staggering and can be used for global (Figure 13a) and regional (Figure 13b) applications. Compared to traditional grid nesting, Voronoi meshes can cleanly incorporate both downscaling and upscaling effects. It can also handle variable resolution at any region of interest even using other grid shapes. For example, Figure 14 shows selective mesh refinement based on terrain height. Note that the high terrains are accompanied by higher resolutions.

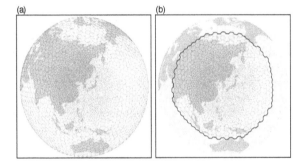

Figure 13. Variable resolution mesh at (a) global and (b) regional scale [32].

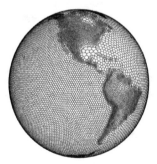

Figure 14. Selective mesh refinement based on centroidal Voronoi terrain height (courtesy Michael Duda).

5.4. Parametarization and grid-spacing

Parameterizations approximate the combined effects of physical processes (e.g., cumulus convection, radiation, microphysics, planetary boundary layer) that are too small, too complex, or too poorly understood to be explicitly represented in numerical models. Models with a grid spacing of 10 km or larger, which is roughly 10 to 20 times the size of the cumulus cloud, needs much finer resolution to resolve small cumulus clouds well. Parameterization schemes in a model are often optimized for a certain range of grid-spacing in the model. Such dependence on the grid-spacing makes a parameterization not suitable when the grid-spacing of the model decreases or increases. Most mesoscale models that can be used over a wide range of grid-spacing typically have multiple parameterization schemes for the same process to be used at multiple resolutions. Such dependence of parameterization schemes on the grid-spacing often works as an obstacle towards the use of higher resolution.

6. Summary

Since the first forecast, the complexity and sophistication of numerical weather prediction models have increased tremendously [35]. The continued improvement in data assimilation and numerical models, and the continued availability of ever larger and faster computers have allowed numerical weather prediction to become more accurate [36]. Most of the improvements in numerical models that occurred over the past 50 years can be categorized as either improved numerical techniques, improved model resolution, or improved model physics. Development of different types of grids, grid-staggering and grid-spacing are intimately related to the improvement in the weather and climate prediction.

There are three major shapes of grids, namely latitude-longitude, triangular and hexagonal. Although latitude-longitude grid is the most common, it suffers from the polar problem. This problem can be overcome by using a triangular or hexagonal grid. Variables in a grid can be defined at the same point ("unstaggered grid") or at different points ("staggered grid"). The staggered grid often has higher effective resolution than the unstaggered grid and is used in almost all popular models. Staggering can also be implemented in the vertical direction and in time. Physics parameterization schemes in a model are often optimized for a certain range of grid-spacing. Such dependence on the grid-spacing makes a parameterization not suitable when the grid-spacing of the model decreases or increases. Further development of the grids is needed to better represent the variables in a model for weather and climate prediction.

Author details

Sarah N Collins, Robert S James, Pallav Ray*, Katherine Chen, Angie Lassman and James Brownlee

Department of Marine and Environmental Systems, Florida Institute of Technology, Melbourne, FL, USA

References

[1] Stensrud, D. J. Parameterization Schemes Keys to Understanding Numerical Weather Prediction Models. Cambridge University Press, (2007).

[2] Duran, D. R. Numerical Methods for Fluid Dynamics. Texts in Applied Mathematics, (2010).

[3] Gates WL, AMIP: The Atmospheric Model Intercomparison Project. Bull. Amer. Meteor. Soc. 1992;73, 1962-1970

[4] Swinbank, R. Numerical Weather Prediction. Data Assimilation, Springer Berlin Heidelberg, (2010). , 381-406.

[5] Kwizak, M, & Robert, A. J. A Semi-Implicit Scheme for Grid Point Atmospheric Models of the Primitive Equations. Mon. Wea. Rev., (1971). , 32-36.

[6] Bjerknes, V. Das Problem der Wettervorhersage, Betrachtet vom Stanpunkt der Mechanik und der Physik. Meteor. Zeits, (1904). , 1-7.

[7] Abbe, C. The Physical Basis of Long-Range Weather Forecasts. Mon. Wea. Rev., (1901). , 551-561.

[8] Richardson, L. F. Weather Prediction by Numerical Process. Second Edition, Cambridge University Press, (2007).

[9] Stewart, R. Numerical Weather Prediction. Weather forecasting by Computer. (2008). Available online at http://www.robinstewart.com/personal/learn/wfbc/numerical.html#_ftn4.

[10] Burridge, D. M. Some Aspects of Large Scale Numerical Modeling of the Atmosphere. Proceedings of 1979 ECMWF Seminar on Dynamical Meteorology and Numerical Weather Prediction, (1980). , 2, 1-78.

[11] Randall, D. A, Ringler, T. D, Heikes, R. P, Jones, P, & Baumgardner, J. Climate Modeling with Spherical Geodesic Grids. Computing in Science & Engineering, (2002). , 32-40.

[12] Kurihara, Y. Numerical Integration of the Primitive Equations on Primitive Grids. Mon. Wea. Rev., (1965). , 399-415.

[13] Puser, R. J. Accurate Numerical Differencing Near a Polar Singularity of a Skipped Grid. Mon. Wea. Rev., (1988). , 1067-1076.

[14] Rancic, M, & Zhang, H. Variable Resolution of Quasi-Uniform Grids: Linear Advection Experiments. Meteor. Atmos. Phys., (2006). doi:s00703-005-0165-4

[15] Sadourney, R, Akio, A, & Yale, M. Integration of the Non-Divergent Barotropic Vorticity Equation with an Icosahedral-Hexagonal Grid for the Sphere. Mon. Wea. Rev. (1968). , 351-356.

[16] Majewski, D, Dorte, L, Peter, P, Bodo, R, Michael, B, Thomas, H, Gerhard, P, & Werner, W. Operational Global Icosahedral-Hexagonal Gridpoint Model GME: Description and High-Resolution Tests. Mon. Wea. Rev., (2002). , 319-338.

[17] Thuburn, J. A PV-Based Shallow Water Model on a Hexagonal-Icosahedral Grid. Mon. Wea. Rev., (1997). , 2328-2347.

[18] Arakawa, A, & Lamb, V. R. Methods of Computational Physics. New York: Academic Press, (1977). , 173-265.

[19] Mesinger, F. Horizontal Advection Schemes on a Staggered Grid, an Enstrophy and Energy Conserving Model. Mon. Wea. Rev., (1981). , 467-478.

[20] Cullen MJPand Davies T. A Conservative Split-Explicit Integration Scheme With Fourth-Order Horizontal Advection. Quart. J. Roy. Meteor. Soc., (1991). , 993-1002.

[21] Randall, D. A. Geostrophic Adjustment and the Finite-Difference Shallow-Water Equations. Mon. Wea. Rev., (1994). , 1371-1377.

[22] Phillips, N. A. The Nested Grid Model, NOAA Technical Report NWS 22. Dept. of Commerce, Silver Spring, MD, (1979).

[23] Janjic, Z. I. Nonlinear Advection Schemes and Energy Cascade on Semi-Staggered Grids. Mon. Wea. Rev., (1984). , 1234-1245.

[24] Lorenz, E. N. Energy and Numerical Weather Prediction. Tellus, (1960). , 157-167.

[25] Arakawa, A, & Moorthi, S. Baroclinic Instability in Vertically Discrete Systems. J. Atmos. Sci., (1988). , 1688-1707.

[26] Charney, J. G, & Phillips, N. A. Numerical Integration of the Quasi-Geostrophic Equations for Barotropic and Simple Baroclinic flows. J. Meteor., (1953). , 71-99.

[27] Arakawa, A. Adjustment Mechanisms in Atmospheric Motions. J. Meteor. Soc. Japan, Special issue of collected papers, (1997). , 155-179.

[28] Skamarock, W. C, Klemp, J. B, Dudhia, J, Gill, D. O, Baker, D. M, & Duda, M. G. Huang XYu., Wang W., Power JG. A Description of the Advanced Research WRF Version 3. NCAR/TN-475+STR, (2008). , 125.

[29] Eliassen, A. A Procedure for Numerical Integration of the Primitive Equations of the Two-Parameter Model of the Atmosphere. Science report 4, Department of meteorology, UCLA, (1956).

[30] Bratseth, A. Some Economical Explicit Finite-Difference Schemes for the Primitive Equations. Mon. Wea. Rev., (1983). , 663-668.

[31] Mesinger, F. A Method for Construction of Second-Order Accuracy Difference Schemes Permitting no False Two-Grid Interval Wave in the Height Field. Tellus, (1973). , 25, 444-458.

[32] Model for Prediction Across Scale (MPAS)Regional Training Workshop on WRF, Sep (2012). Hanoi, Vietnam., 24-28.

[33] Ringler, T, Ju, L, & Gunzburger, M. A Multiresolution Method for Climate System Modeling: Application of Spherical Centroidal Voronoi Tessellations. Ocean Dynamics, (2008). , 475-498.

[34] Skamarock, W. C, Klemp, J. B, Duda, M. G, Fowler, L. D, & Park, S. H. A Multiscale Nonhydrostatic Atmospheric Model Using Centroidal Voronoi Tesselations and C-grid Staggering. Mon. Wea. Rev., (2012). , 3090-3105.

[35] Lynch, P. The Origins of Computer Weather Prediction and Climate Modeling. Journal of Computational Physics, (2008). , 3431-3444.

[36] Kalnay, E. Atmospheric Modeling, Data Assimilation and Predictability. Cambridge University Press, (2003).

Chapter 5

Impact of Tropical Cyclone on Regional Climate Modeling over East Asia in Summer and the Effect of Lateral Boundary Scheme

Zhong Zhong, Yijia Hu, Xiaodan Wang and Wei Lu

Additional information is available at the end of the chapter

1. Introduction

Regional climate models (RCMs) have been increasingly used to simulate the realistic characteristics of the regional climate and to evaluate the related multiscale interactions [1-4]. A fundamental requirement for the development of RCMs is to examine their capability in simulating the seasonal evolution and interannual variabilities on a regional scale. This seems more important for the simulation of RCMs over East Asia due to the distinct Asian monsoon climate feature. The larger variability and uncertainty of the Asian monsoon system confirmed by extensive observational studies and numerical simulations [5, 6] suggest that the modeling research will be a challenging task because of the complex scale interactions involved.

China is strongly influenced by the Asian monsoon circulation. The complexity and regional diversities of the summer precipitation over China is a universally accepted understanding in literatures. Because of the complex terrain and great variability of the monsoon climate in East Asia, simulation of the East Asian climate is sensitive to physical schemes [7, 8] and the skill is still limited [9-10]. One of the main atmospheric systems affecting the summer precipitation over China is the western Pacific subtropical high (WPSH). The seasonal extension and withdrawal of the WPSH are responsible for the rainfall distribution over eastern China. Therefore, the simulated precipitation is directly dependent on the capability of the models to capture the activities of the WPSH and the corresponding regional circulations. In general, both RCMs and general circulation models (GCMs) can reproduce the seasonal evolution of the WPSH [11]. However, the simulated high in summertime is either weaker [12] or stronger [13] than the observed one depending on the observational driven fields and the specific monsoon case. On the other hand, it is more difficult to simulate reasonably well the behavior of the WPSH over a shorter time scale during the summer monsoon period, which plays a significant role in the simulation of the precipitation over eastern China [14].

For an East Asian summertime case in 1994, the seasonal scale simulations using the Regional Climate Model version 3 (RegCM3), developed by the International Centre for Theoretical Physics (ICTP), was conducted with three cumulus parameterization schemes (CPSs), i.e., the Emanuel scheme [15], Grell scheme [16] and Kuo scheme [17]. It was found that the model is capable of reproducing the mean summer circulation over East Asia, as has been demonstrated by previous works [12, 13, 18-21]. However, the model shows a failure simulation over shorter time scale. Figure 1 depicts the temporal evolution of the spatial abnormal correlation coefficients (ACCs) for geopotential height of 500 hPa between the observation and simulations. The averaged ACCs in Figure 1 are 0.89, 0.88 and 0.84 corresponding to different CPS, and the temporal tendency of the ACCs is basically similar in the three experiments. However, the evolution of the ACCs exhibits a sharp oscillation with three lower-value periods. This suggests that the model can reasonably reproduce the regional circulation pattern with all the three CPSs in most of the periods, but it yields some unsuccessful simulations in the cases with lower ACCs. What kind of circulation systems cannot be reproduced well by the model in the periods with lower ACCs? To gain insight into the possible cause for these lower ACCs, the selected four periods shown in Figure 1, i.e., June 8–12 (period A), July 18–22 (period B), July 24–28 (period C) and August 27–31 (period D), are analyzed in detail.

The mean geopotential height of 500 hPa and the wind field for the observation and simulation with three CPSs during period B (July 18–22) are shown in Figure 2. In this period, when the strong WPSH dominates over the western North Pacific (WNP) and the southwest and south monsoon flow prevail over eastern China, the mean circulation system over the eastern part of the model domain is evidently different from that in summertime. The observed circulation clearly shows that the WPSH is interrupted by a TC and the main body of the WPSH withdraws eastward. The cyclonic circulation system holds over the WNP, whereas a split anticyclonic cell from the main body of the WPSH is over eastern China and the southern Korean Peninsula (Figure 2a). Obviously, the model with all the three CPSs cannot suitably reproduce the invading process of the TC and the associated split of the WPSH due to the deficient simulations of the TC activity (Figures 2b–2d).

Moreover, it can be seen that the simulated TC with the Emanuel and Grell schemes is stronger and turns northeastward in advance over the WNP to the east of Japan, whereas the simulation with the Kuo scheme gives a lower pressure system over Japan. Over the WNP where the observed TC is located, the anticyclonic circulation appears clearly. The similar situation also occurred for period C and period D in Figure 1 when TC was over WNP. Therefore, for the amelioration of RCMs, the important problem of model development on how to enhance the capacity for portraying the activity of TCs in summertime over the WNP should be solved.

On the contrary, during period A with higher ACCs as shown in Figure 1, the regional circulation systems are suitably portrayed in the model because this period is in the intermittent period of the TCs in the model domain (Figure 3). This clearly indicates that the simulated trough and ridge in the middle and high latitudes resemble the observed ones. The simulated WPSH in the lower latitude is also largely consistent with the observation, although the simulated one seems

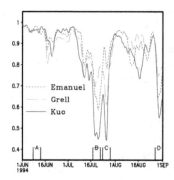

Figure 1. Temporal evolution of spatial abnormal correlation coefficients (ACCs) of 500hPa geopotential height with CPSs of Emanual scheme, Grell scheme and Kuo scheme, respectively (A represents the period with higher ACC, whereas B, C and D represents the period with lower ACC, respectively).

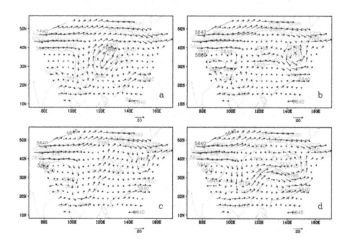

Figure 2. The temporal mean geopotential height and wind vector at 500hPa observed (a) and simulated with Emanuel scheme (b), Grell scheme (c) and Kuo scheme (d), respectively, during period B in Figure 1.

slightly weaker, with each of the three individual CPSs. Overall, the simulations with the Emanuel scheme or Grell scheme exhibits better results than that with Kuo scheme; the latter generates a weakened WPSH and an abnormal split trough in the middle latitude, which implies that the regional climate modeling is sensitive to model physics, such as CPS.

In spite of the examples of the tropical storms as [18], the model can generally reproduce their intensity and track, the regional climate modeling over East Asia in summertime on the impact of bogus typhoons implies that the model cannot suitably reproduce the impact of TCs without a special bogus technique [22]. TCs hinder the precise simulation of the summer monsoon circulation over East Asia [23]. It is a challenge for regional climate modeling to overcome the

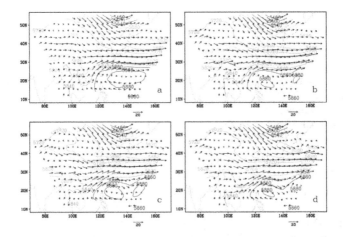

Figure 3. Same as figure 2, except for the period A shown in Figure 1.

disadvantage of the failure simulation when TCs are active over the WNP for the simulation over East Asia. In addition to the TC bogus technique, which is complicated for long-term simulation, improvements in both the treatment for lateral boundary conditions and model physical parameterizations could be helpful to more suitably represent the behavior of TCs in the model domain. The possible improvement may include selecting the proper buffer zone size, adopting different nudging coefficients in the lower and upper troposphere, etc. With the amelioration, the TCs over the WNP would be regulated in the right manner, even though the climate model at a coarser resolution cannot describe the detailed structure and intensity of TCs.

2. Effects of lateral boundary scheme on the TC track simulation in RCM

The lateral boundary condition (LBC) is one of the basic issues of the RCM, generally, a good performance of the RCM is dependent to the good LBCs. The success of the RCM depends on the adoption of appropriate lateral boundary technique [2, 24, 25]. Previous studies have shown that it is more important to provide a good lateral boundary condition for the RCM than to improve the physical process schemes. A broader buffer zone size (BZS) had been recommended for the simulation of the East Asian summer monsoon using the RCM [26]. Majority of RCMs developed to date use the so-called nudging technique as in [27] to develop a meteorological LBC that involves the application of a Newtonian term and a Laplace diffusion term to drive the model solution toward large-scale driving fields over the lateral boundary buffer zones, in insuring a smooth transition between LBC-dominated and model-dominated regimes and in reducing noise generation [28]. As the climate of the regional model is the equilibrium of the atmospheric physical and dynamical processes and the information provided by the LBCs [2], then, whether or not the lateral boundary scheme affects the track of the TC in the model domain is a technical problem of the RCM [29].

Tropical cyclone Winnie (1997), which developed over the central Pacific on August 8, 1997, was the strongest TC in the Western Pacific in that year. It brought considerable loss of life and property and adversely affected China's national economy because Winnie passed through eastern China after its landfall at Wenling, Zhejiang Province, on August 18. Taking August of 1997 as an example and using RegCM3, we will examine the impact of lateral boundary buffer zone scheme on the ability of the RCM to describe TC activity in an effort to provide a basis for improvement of the simulation of the East Asian summer monsoon climate.

2.1. Model description and experimental design

RegCM3 is an upgraded version of the model originally developed by Giorgi et al. [28, 30] and then improved upon as discussed by Giorgi and Mearns [2] and Pal et al. [31]. The dynamic core of RegCM3 is equivalent to the hydrostatic version of the NCAR/Pennsylvania State University mesoscale model, MM5 [32]. The physical parameterizations employed in this simulation include the comprehensive radiative transfer package of the NCAR Community Climate Model CCM3 [33], the nonlocal boundary layer scheme of Holtslag et al. [34], the BATS land surface model [35], and the cumulus parameterization scheme of Grell with the Fritsch-Chappell-type closures [32]. As shown in Figure 4, the model domain is centered at (32.5°N, 120°E) with 121 east-west points and 80 north-south points and a horizontal grid spacing of 60 km. The top of the model is at 50 hPa with 18 uneven levels in the vertical and the buffer zone size is 14 grid-point width.

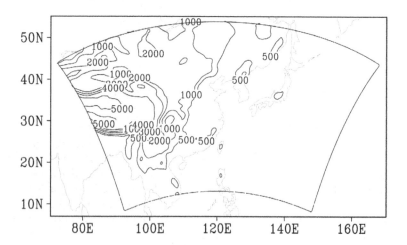

Figure 4. Model domain and topography (units: m).

RegCM3 adopts nudging technology to introduce the large-scale forcing in the buffer zone of lateral boundaries. For a variable α, the nudging equation in the buffer zone can be written as

$$\frac{\partial \alpha}{\partial t}(n) = F(n)F_1(\alpha_L - \alpha_M) - F(n)F_2\nabla^2(\alpha_L - \alpha_M) \tag{1}$$

In equation (1), subscripts L and M refer to the large-scale forcing field and model simulation field, respectively. $F(n)$ is a function of the buffer zone ordinal n. $F_1 = \frac{a}{\Delta t}$ and $F_2 = \frac{(\Delta s)^2}{b\Delta t}$. Nudging parameters a and b, which has a default value of 0.1 and 50, respectively, can be adjusted to make a strong nudging (approaching to forcing fields rapidly) or weak nudging (approaching to forcing fields slowly). Δt and Δs are the time step and horizontal grid spacing of the model, respectively. A total of 25 experiments were conducted to address the nudging parameters of the model prognostic equations within the buffer zone, where a is set at 0.05, 0.075, 0.1, 0.15 and 0.2 and b at 100, 66.7, 50, 33.3 and 25, respectively. Table 1 lists the experimental names and the corresponding configurations of nudging parameters a and b.

name	NP01	NP02	NP03	NP04	NP05
a / b	0.05/100	0.05/66.7	0.05/50	0.05/33.3	0.05/25
name	NP06	NP07	NP08	NP09	NP10
a / b	0.075/100	0.075/66.7	0.075/50	0.075/33.3	0.075/25
name	NP11	NP12	NP13	NP14	NP15
a / b	0.10/100	0.10/66.7	0.10/50	0.10/33.3	0.10/25
name	NP16	NP17	NP18	NP19	NP20
a / b	0.15/100	0.15/66.7	0.15/50	0.15/33.3	0.15/25
name	NP21	NP22	NP23	NP24	NP25
a / b	0.20/100	0.20/66.7	0.20/50	0.20/33.3	0.20/25

Table 1. Experimental names corresponding to the configuration of nudging parameter a and b

The model was employed to conduct the 1 month long simulation for August 1997, examining the effect of adjusting nudging parameters a and b in the buffer zone on TC track simulation. Each experiment starts at 00:00 GMT on August 1 and ends at 18:00 GMT on August 31. The initial and LBCs were provided by National Centres for Environmental Prediction/National Centre for Atmospheric Research (NCEP/NCAR) reanalysis data, and the LBCs were updated at 6 h interval. Sea surface temperature data are taken from the Global Ocean Surface Temperature (GISST) of the Hadley Center and updated once a week. In addition, the daily precipitation from Global Precipitation Climatology Project (GPCP) data at 1° resolution was used to evaluate the precipitation distribution during the simulation period. The integration time step is 180 s. We specifically analyze the simulation result for Winnie in the model region from August 12 to August 20, whereas the simulations for the first 11 days are not considered to allow for the spin-up of the model [36].

2.2. Effects of nudging parameters

2.2.1. The track of Winnie

Figure 5 presents the simulated Winnie tracks for 25 experiments with different configuration of the nudging parameters a and b. It clearly shows that the nudging parameter has a great impact on the simulation of Winnie track, and an appropriate configuration can effectively improve the track simulation. Among all experiments, NP19 ($a=$ 0.15 and $b=$ 33.3) shows its well performance for the westward track of Winnie entering the East China Sea, though the track error is still significant.

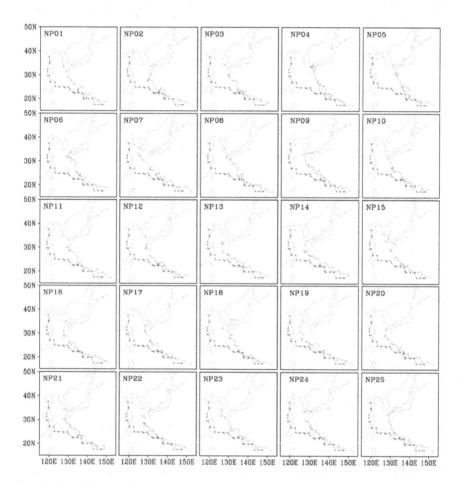

Figure 5. Observed (filled circle) and simulated (open circle) tracks of Winnie at 6 h interval from 00:00 GMT on August 12 to 12:00 GMT on August 20, 1997, for 25 experiments with different configurations of nudging parameters.

However, an inappropriate configuration results in a much greater error in simulating the track; accompanying with a more obvious turning ahead of time and landfall on southern Japan. The RMSEs of 25 experiments are listed in Table 2, which also presents that NP19 performed best among all experiments with different nudging parameter configurations, and its RMSE is less than half of the NP05. Therefore, although the problem of the turning ahead of time in the simulation cannot be solved fundamentally by configuring the nudging parameters, an appropriate configuration can effectively reduce the simulation error of TC track. On the other hand, It can also be seen from Table 2 that the ratio of the two nudging parameters a and b is an important reference in the configuration of nudging parameter. It suggests that if a smaller value is selected for a, then a larger value must be selected for b, and vice versa, which implies that, within the buffer zone, two additional terms with equilibrium relationship in nudging equation (1) are necessary for the better performance of the model. In contrast, the track simulation would be suboptimal. Furthermore, the default setting of the two nudging parameters of the model keeps the equilibrium relationship. When the equilibrium relationship is achieved, either strong nudging experiment (a is big and b is small) or weak nudging experiment (a is small and b is large) is beneficial for the simulation of the Winnie track. When a or b exceeds the value range in Table 1, the model integration would be overflow, which is in accordance with that proposed by Marbaix et al. [37].

a					
b	25	33.3	50	66.7	100
0.05	512.6	436.1	367.8	409.0	334.9
0.075	377.6	327.9	365.8	276.8	412.3
0.1	357.1	309.8	297.6	332.8	374.7
0.15	308.6	248.6	418.9	366.2	346.1
0.2	389.0	360.3	303.6	353.6	339.8

Table 2. RMSE between the observed and simulated track of Winnie for the experiment with different configuration of nudging parameter a and b

2.2.2. The intensity of WPSH

Figure 6 shows the observed and simulated temporal variation of the intensity index of WPSH (IWPSH), defined as the number of grid points with a geopotential height being greater than 5880 gpm in the area east of 110°E, for the 25 experiments configuring with different nudging parameters. It is shown that although the simulated evolution of the intensity index differs for different parameter configuration in detail, the basic feature of the evolution for each experiment is not substantially different. The correlation coefficients for the time series of intensity index of WPSH between observation and simulation experiment from NP01 to NP25 is 0.8168, 0.7964, 0.8495, 0.7813, 0.8038, 0.8098, 0.8504, 0.8365, 0.8270, 0.8455, 0.8106, 0.8411, 0.7873, 0.8227, 0.8615, 0.7975, 0.8331, 0.8554, 0.9027, 0.8239, 0.8382, 0.8350, 0.8957, 0.8641 and 0.8080, respectively. Here again, the NP19 obtains the best performance for the temporal variation of the intensity index of WPSH, meanwhile, the simulated intensity index shows its variation in well agreement with the observed one before 12:00 GMT of August 14, and the weak degree of the simulated WPSH is smallest after August 17 among all 25 experiments.

Furthermore, the sharply variation of the intensity index on August 17 can not be captured also for all 25 experiments with different configuration of nudging parameters, and the simulated intensity index transition appears on August 16. In addition, all the experiments show their higher intensity index before 12:00 GMT on August 14, and lower intensity index after 00:00 GMT on August 15 than that of the observation. Therefore, it could be concluded that no experiment with different configuration of nudging parameters could solve the problem of turning of the Winnie ahead of time fundamentally. But an appropriate configuration can be effective insuring the simulated track more being close to the observed one.

From the simulated pattern of the precipitation rate averaged between August 12 to August 19, one can see that although an appropriate nudging parameter configuration could partly improve the precipitation pattern simulation, it does not make the simulation being consistency with the observation entirely, which implies that the description of the precipitation physics in the model is still vital in simulating the precipitation pattern (figures are not shown).

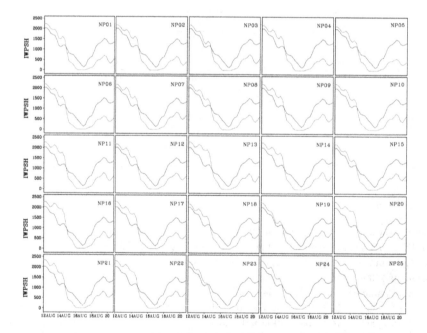

Figure 6. Temporal variations of the observed (solid line) and simulated (dotted line) intensity indices of the western Pacific subtropical high from 00:00 GMT on August 12 to 12:00 GMT on August 20, 1997, for 25 experiments with different configurations of nudging parameters.

To verify the importance of TC on the regional circulation simulation, we depicted the observed and NP19-simulated daily geopotential height at 500 hPa on August 12 and August 16, respectively (Figure 7). It can be seen that the mean circulation over the model domain at 500 hPa on August 12 was reproduced well, for the well-simulated position of Winnie (Figure 7a and 7b). However, the discrepancy between the simulated and observed circulation south to 35°N is much greater for about 5 degree longitudes error in position and 120 gpm error in the intensity of Winnie on August 16 (Figure 7c and 7d). The correlation coefficients between observations and simulations for the region east to 100°E and south to 35°N is 0.9694 and 0.7315, respectively. Therefore, the evolution of 500 hPa geopotential height, as well as the intensity of WPSH, is closely related to the TC track, and the simulated TC track error is the primary cause for the failure in simulating east Asian climate in summer [8].

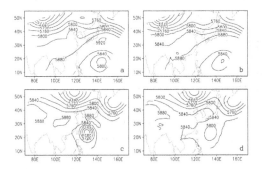

Figure 7. Observed (a, c) and NP19-simulated (b, d) daily geopotential height at 500 hPa on August 12 (a, b) and August 16 (c, d), respectively.

3. Impacts of tropical cyclones on the regional climate over East Asia

It is well known that during summer, WPSH is the predominant large-scale circulation system over the WNP. In most cases, this circulation system regulates the tracks of the TCs. However, the TCs in turn influence the WPSH as well. When a stronger TC turns northeastward over the WNP, it usually causes the WPSH to split [8, 38]. For operational prediction, the impacts of circulation systems, particularly the WPSH, on the tracks of TCs have been studied in detail. However, the quantitative effect of TCs on the regional circulation remains unknown.

Due to the limited observational data available, it is difficult to assess the quantitative effects of TCs on regional circulation systems, particularly on the state of regional circulation without TCs. The RCM is an effective tool that can be used to address this issue. It was noted that the climate in a regional model is determined by a dynamical equilibrium between two factors, i.e., the information provided by the lateral boundary condition and the internal model physics and dynamics [12]. This suggests that the performance of the regional model is mostly dependent on the lateral boundary condition [39, 40]. From this view point, the effects of TC on regional circulation systems can be evaluated in terms of the removal of TCs at the lateral boundaries of the model. Therefore, the circulation systems over the model domain would be unaffected by the TCs. In this section, as a case study, RegCM3 is used to determine the extent to which the regional climate is affected by the TCs. This was done by comparing the simulation results at the climate scale with and without TCs at the lateral boundaries of the model's driven fields. The simulation begins at 0000 GMT on July 15 and ends at 1800 GMT on August 31, 1997. During the week of August 17–23 in the simulation period, China suffered huge economic losses due to the damage caused by the violent TC Winnie (1997).

Two experiments were performed, one was the control run (CR) as described above, and the other was the sensitivity run (SR). The SR was conducted by removing the TCs from the 6h interval large-scale driven fields for the same period as that of the CR. This strategy is in agreement with the removal technique of the large-scale TC circulation from the first-guess fields before the bogus TC is inserted into the initial fields, commonly used for the numerical simulation or prediction of TCs [41]. In our approach, we modified the vorticity, geostrophic vorticity and divergence. Then, we solve for the change in the nondivergent stream function, geopotential and velocity potential, and compute the modified velocity field, temperature field and the corresponding geopotential height field. The details of the strategy employed for the removal of TCs from large-scale driven fields can be found in [41].

Figure 8 shows the monthly geopotential height and wind vector of the observations, CR and SR at 200 hPa, 500 hPa and 850 hPa in August 1997. It is observed that the CR performs well for the regional circulation in the lower troposphere over East Asia, while the East Asian summer monsoon predominates the southeastern coast of China and the adjacent WNP; further, a cyclonic circulation interrupts the southwestern summer monsoon over south China (bottom panels in Figures 8a and 8b). The simulated ridge line of WPSH at 500 hPa is at 30°N; this is consistent with the observations (middle panels in Figures 8a and 8b). At 200 hPa, the mean circulation pattern and the South Asia High (SAH) are also reproduced well (top panels in Figures 8a and 8b). However, at 850 hPa, the simulated WPSH is slightly weaker, whereas the lower depression system over the west part of northeastern China is stronger than the observed one, which is partially caused by the TC Winnie passed through Bohai Sea and landed again at Yinkou, Liaoning province, and activated over northeast China.

It is noteworthy that the mean circulation in the lower troposphere in August 1997 is somewhat different from the normal summer monsoon pattern while the southwest summer monsoon flow prevails over south China [13]. The cyclonic circulation at 850 hPa, which interrupted the southwest summer monsoon over south China in August 1997 in the mean chart, was mainly caused by the landfall of TCs, particularly that of the violent TC Winnie on 19 August and its sweep over eastern China in the subsequent days.

In the case of the SR, the simulated WPSH in the lower troposphere intensifies and extends westward significantly. Moreover, the summer monsoon in the lower troposphere with a stronger southerly component from the South China Sea and the Bay of Bengal is predominant over the southeast mainland of China. Meanwhile, the simulated intensity of the lower depression system recovered for TC Winnie is no longer active and finally filling up over northeastern China (shown in the middle and bottom panels in Figures 8c). The simulated

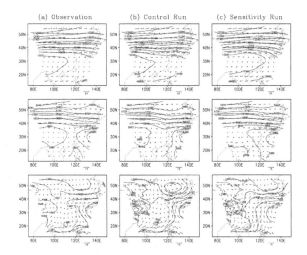

Figure 8. Monthly geopotential height (contour lines) and wind (vector arrows) of the observation (a), CR (b) and SR (c) at 200 hPa (top panel), 500 hPa (middle panel) and 850 hPa (bottom panel) in August 1997.

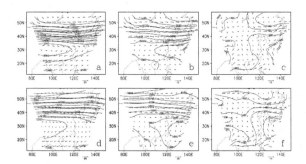

Figure 9. Monthly geopotential height (contour lines) and wind (vector arrows) at 200 hPa (a), 500 hPa (b) and 850 hPa (c) in August 2003.

SAH at 200 hPa is also intensified. However, it extends eastward distinctly, thereby confirming that the WPSH and SAH act in the opposite directions, as noted by Wu et al. [38]. The simulated mean circulation pattern for SR can be verified by the observations. As an example, Figure 9 shows the monthly circulation at 200 hPa, 500 hPa and 850 hPa in August 2003, when TCs over the WNP are observed to be less than usual in this month. It also exhibits a similar pattern for the mean circulation for SR, while the SAH is intensified and extends eastward (Figure 9a); further, the WPSH is also intensified, but it extends westward (Figure 9b), and the southerly monsoon flow prevails over Central and Southern China in the lower troposphere (Figure 9c).

The difference in monthly circulations between SR and CR are shown in Figure 10. It clearly shows an anticyclonic circulation in the entire troposphere over eastern China and the adjacent

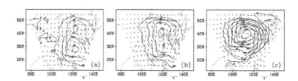

Figure 10. Monthly differences in the geopotential height (contour lines) and wind (vector arrows) between SR and CR at 850 hPa (a), 500 hPa (b) and 200 hPa (c).

WNP. Since TC Winnie was the most violent one in 1997, it significantly influences the difference circulation chart and it exhibits an anticyclonic difference circulation in the middle and lower troposphere and a circular anticyclonic difference circulation at 200 hPa. Moreover, the axis of anticyclonic difference circulation in the vertical is inclined westward above 500 hPa centred at approximately 120°E in the lower troposhpere and over the bend of the Yellow River (40°N, 112°E) at 200 hPa.

In order to consider the impacts of TCs on the strength of the WPSH, the intensity index of the WPSH (IWPSH) defined in section 2 is employed here to assess the effects of the TCs on the strength of the WPSH quantitatively. Figure 11 shows the temporal variations of IWPSH in the observation, CR and SR. Four major intensification processes are observed on around August 2, 11, 22 and 30, and three weakening processes are observed on around August 7, 17 and 25. When the activities of the TCs in this period are compared, it is found that each weakening process is attributed to the presence of the following TC over the WNP: Tina, Winnie and Amber, respectively. As shown in Figure 12, these TCs occurred over the WNP during three weakening process respectively and they occupied the position where the WPSH was originally located, resulting in the lower IWPSH. However, the three intensification processes (on August 2, 22 and 30, respectively) are attributed to the presence of the following TCs over the southern coastal area of China and to the west of the WPSH: Victor, Zita and Amber (as shown in Figure 13a, 13c and 13d). These are beneficial to the strengthening of the WPSH over the ocean. Another intensification process (on around August 11) occurred during the intermittent period of TC over the WNP (Figure 13b). This reduction in intensity when the TCs are active over the WNP demonstrates the influence of TCs on the WPSH. Here, it could be inferred that the effect of monsoon latent heating on the strength of subtropical anticyclones, as emphasized by Hoskins [42] and Liu et al. [43], may play a secondary role in the evolution of WPSH and it will be effective only if the TCs are not active over the WNP. Meanwhile, the latent heat release in the cloud wall around the centre of the TCs over southeastern China may have the same effect as that of monsoon latent heating, which will generate a secondary vertical circulation with an ascending branch in the cloud wall and a descending one in the WPSH. This could be the reason why the WPSH is generally intensified after a TC landfall occurs over China. A specific example is TC Amber, which weakened the WPSH when it was over the WNP, but intensified the WPSH after its landfall over China. In fact, TC Winnie also exhibited the same effect before and after its landfall; it contributed to the intensification process of the WPSH before August 21 when it was over east China.

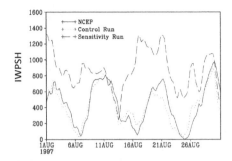

Figure 11. Temporal variation in the intensity index of the WPSH for the observation (solid line), CR (dotted line) and SR (dashed line).

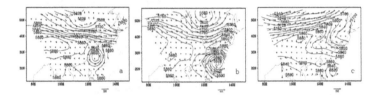

Figure 12. Geopotential height (contour lines) and wind (vector arrows) at 500 hPa on August 7 (a), 17 (b) and 25 (c).

The evolution of the IWPSH for CR shows that the model can reproduce the temporal variation of the WPSH intensity, which demonstrates that the model performs well for the variation in circulation over WNP as well as for the activities of TCs. However, in SR, the simulated evolution of the IWPSH is different from that in the observed one and CR. The most distinct feature is that the three intensification processes (on August 2, 22 and 30) are magnified except for the one on around August 11, which is not directly related to TC. On the other hand, an abnormal intensification process is observed during August 14–22 for the WPSH simulation when TC Winnie is removed from the large-scale forcing fields at the eastern lateral boundary of the model. It should be noted that the mechanisms of the observed and SR-simulated intensification processes of WPSH in those periods are different, i.e., the effects of the TCs are included in the observation but they are removed in the SR simulation. Moreover, it is revealed that the model cannot reproduce the intensification of the WPSH on around August 11, because it is not related to the activity of the TC over WNP.

Generally, TCs over the WNP or China result in substantial precipitation over central and eastern China, which can mitigate the torridity and drought in East Asia when WPSH dominates over there. However, the TCs will interrupt the extension of the East Asian summer monsoon on the weather and climate scales in the lower troposphere, as shown in Figure 8a and 8b. This will eliminate the transportation of water vapour from the Indian Ocean and South China Sea towards eastern China. Figure 14 shows the monthly differences of the precipitable water vapour (PWV) and the percentage rate below 500 hPa between SR and CR. As expected,

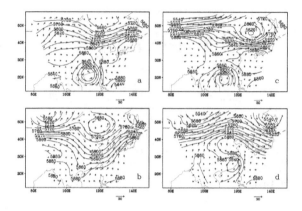

Figure 13. Geopotential height (contour lines) and wind (vector arrows) at 500 hPa on August 2 (a), 11 (b), 22 (c) and 30 (d).

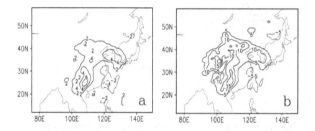

Figure 14. Monthly differences in the precipitable water vapour (a, mm) and the corresponding percentage rate (b) below 500 hPa between SR and CR.

when the TCs are removed, the atmosphere in the lower troposphere over southeast China, the adjacent WNP as well as northeast China would become drier, while that over southwest China, the eastern part of northwest China and north China would become wetter.

4. Conclusions and discussions

As a case study, RegCM3 was employed to assess the possible failure of regional climate model in simulating East Asian summer monsoon. It was found that the model does not perform well when TCs are active over the WNP and it exhibits a common feature when TCs are over WNP in model domain, which gives an anticyclonic circulation difference over WNP, and two cyclonic circulation differences centered over eastern China/Yellow Sea and Japan/WNP to the east of Japan, depending on the adopted CPS. With a smaller model domain, the model can give a more reasonable result over East Asia forced by reanalysis data.

TCs over the WNP and China make great contribution to the formation of the regional climate over East Asia, and the regional climate is significantly affected by the frequency of TC activity in terms of weakening the WPSH when TCs are over the WNP and interrupting the summer monsoon when TCs make landfall over China. If there were no TCs, the atmosphere over southeast China and northeast China would become drier while that over southwest China and north China would become wetter.

The configuration of nudging parameters for the model buffer zone can significantly affects the TC track simulation. Although the best configuration of nudging parameters does not completely eliminate the error in simulating the track, it can largely reduce the errors. Different parameter configuration would generate RMSE in the track simulation by more than two times. Therefore, the appropriate nudging parameter configuration, which maintains the equilibrium between Newtonian term and Laplace diffusion term of the prognostic equations in buffer zone, is crucial for improving the TC track simulation. Neither strong nudging experiment nor weak nudging experiment is beneficial for simulation of the TC track.

To some extent, the error in the simulation of the TC track is related to that of the intensity of WPSH. The weaker simulated WPSH would be the cause of the turning of TCs ahead of time, which will result in great error of track simulations against track observations. Therefore, improvement only with the RCM lateral boundary scheme does not fundamentally eliminate errors in the simulation of the TC tracks. The key challenges in eliminating the errors are to determine how to solve the problem of simulated weaker WPSH, as well as the appropriate presentation of the interaction between TCs and WPSH.

Moreover, to compare the impact of resolution on the track simulation, experiments at a horizontal grid spacing of 30 km for the different buffer zone size were also performed. It seems a more reasonable track can be obtained from the simulation at a higher resolution, but the TC is not landed as reality for all the experiments, and the simulated intensity is usually much weaker than observed one (figures are not shown). It should be pointed out that the improvement of track for the simulation at a higher resolution may come from the positioning of TC partially, therefore, the discrepancy of track can not be eliminated totally for the simulation at higher resolution, and meanwhile, one cannot expect the much higher resolution for the climate models, though it has been demonstrated that the model would perform well at high resolution [40] and its performance is also related to the domain choice [44] and buffer zone size [45].

It has been known that the model physics plays a fundamental role in TC simulation, though it is as yet unclear whether and to what degree the simulated TC track, structure, intensifica-tion, and intensity can be affected by using different physics parameterization schemes [46]. For example, the development of TC is sensitive to the transportation of sensible heat, latent heat and momentum in the underlying surface [47], thus the planetary boundary layer scheme will be important in TC simulation. With the MRI mesoscale nonhydrostatic model, it was found that the precipitation structure induced by typhoon Flo is dependent to the microphysics scheme of the model at a great extent [48], which will in turn reflect the track and intensity of TC through feedback and interaction mechanism. In addition, the simulated tracks of TCs are affected by the detailed microphysics transport in the cumulus parameterization scheme [49].

Acknowledgements

This work is supported by the R&D Special Fund for Public Welfare Industry (Meteorology) under Grant No. GYHY201306025 and National Natural Science Foundation of China (41175090)

Author details

Zhong Zhong[*], Yijia Hu, Xiaodan Wang and Wei Lu

*Address all correspondence to: zhong_zhong@yeah.net

College of Meteorology and Oceanography, PLA University of Science and Technology, Nanjing, China

References

[1] Giorgi, F, & Mearns, L O. Approaches to the Simulation of Regional Climate Change: A Review. Review of Geophysics (1991)., 29, 191-216.

[2] Giorgi, F, & Mearns L O. Introduction to Special Section: Regional Climate Modeling Revisited. Journal of Geophysical Research(1999)., 104, 6335-6352.

[3] McGregor, J L. Regional Climate Modeling. Meteorology and Atmospheric Physics (1997)., 63, 105-117.

[4] Wang, Y. Regional climate modeling: Progress, Challenges, and Prospects. Journal of the Meteorological Society of Japan(2004)., 82, 1599-1628.

[5] Ju, J, & Slingo, J. The Asian Summer Monsoon and ENSO. Quartly Journal of Royal Meteorological Society(1995)., 121, 1133–1168.

[6] Lau, K M, & Yang, S. Seasonal Variation, Abrupt Transition, and Intraseasonal Variability Associated with the Asian Monsoon in the GLA GCM. Journal of Climate(1996)., 9, 965-985.

[7] Hong, S-Y, & Choi, J. Sensitivity of the Simulated Regional Climate Circulations over East Asia in 1997 and 1998 Summers to Three Convective Parameterization Schemes. Journal of Korean Meteorological Society(2006)., 42, 361-378.

[8] Zhong, Z. A possible cause of a regional climate models' failure in simulating the East Asian summer monsoon. Geophysical Research Letters(2006)., 33, L24707, doi: 10.1029/2006GL027654.

[9] Zhou, T, Yu, R, Liu, H, & Wang, B. Ocean Forcing to Changes in Global Monsoon Precipitation over the Recent Half-Century. Journal of Climate (2008)., 21, 3833-3852.

[10] Zhou, T, Wu, B, & Wang, B. How Well Do Atmospheric General Circulation Models Capture the Leading Modes of the Interannual Variability of the Asian-Australian Monsoon? Journal of Climate(2009)., 22, 1159-1173.

[11] Wang, Z Z, Wu, G X, Wu, T W, & Yu, R C. Simulation of Asian Monsoon Seasonal Variation with Climate Model R42L9/LASG. Advances in Atmospheric Sciences(2004)., 21, 879-889.

[12] Giorgi, F, Huang, Y, Nishizawa, K, & Fu, C. A Seasonal Cycle Simulation over Eastern Asia and its Sensitivity to Radiative Transfer and Surface Processes. Journal of Geophysical Research(1999)., 104, 6403-6423.

[13] Lee, D-K, & Suh, M-S. The Ten-year East Asian Summer Monsoon Simulation Using a Regional ClimateModel (RegCM2). Journal of Geophysical Research(2000)., 105, 29565-29577.

[14] Zhong, Z, Wang, X, & Min, J. Testing the Influence of WPSH on the Precipitation over Eastern China in Summer Using RegCM3. Theoretical and Applied Climatology(2010)., 100, 67−78.

[15] Emanuel, K A. A Scheme for Representing Cumulus Convection in Large-scale Models. Quartly Journal of Royal Meteorological Society(1991)., 48, 2313-2335.

[16] Grell, G A. Prognostic Evaluation of Assumptions Used by Cumulus Parameterizations. Monthly Weather Review(1993)., 121, 764-787.

[17] Anthes, R A. A Cumulus Parameterization Scheme Utilizing a One-dimensional Cloud Model. Monthly Weather Review(1977)., 117, 1423-1438.

[18] Liu, Y, Giorgi, F, & Washington, W M. Simulation of Summer Monsoon Climate over East Asia with an NCAR Regional Climate Model. Monthly Weather Review(1994)., 122, 2331-2348.

[19] Kato, H, Hirakuchi, H, Nishizawa, K & Giorgi, F. Performance of NCAR RegCM in the Simulation of June and January Climates over Eastern Asia and the High-resolution Effect of the Model. Journal of Geophysical Research(1999)., 104, 6455-6476.

[20] Lee, D-K, Kang, H-S, & Min, K-H. The Role of Ocean Roughness in Regional Climate Modeling: 1994 East Asia Summer Monsoon Case. Journal of the Meteorological Society of Japan (2002)., 80, 171-189.

[21] Wang, Y, Sen O L, & Wang, B. A Highly Resolved Regional Climate Model (IPRC-RegCM) and Its Simulation of the 1998 Severe Precipitation Event over China. Part I: Model Description and Verification of Simulation. Journal of Climate(2003)., 16, 1721-1738.

[22] Lee, D-K, Ahn, Y-I, & Kim, C-J. Impact of Ocean Rough and Bogus Typhoons on Summertime Circulation in a Wave-atmosphere Coupled Regional Climate Model. Journal of Geophysical Research (2004)., dio:10.1029/2003JD003781.

[23] Zhong, Z, & Hu, Y. Impacts of Tropical Cyclones on the Regional Climate: an East Asian Summer Monsoon Case. Atmospheric Science Letters(2007)., 8, 93-99.

[24] Denis, B, Laprise, R, & Caya, D. Downscaling Ability of One-way Nested Regional Climate Models: the Big-Brother Experiment. Climate Dynamics(2002)., 18, 627-646.

[25] Diaconescu, P E, Laprise, R & Sushama, L. The Impact of Lateral Boundary Data Errors on the Simulated Climate of a Nested Regional Climate Model. Climate Dynamics(2007)., 28, 333-350.

[26] Wei, H, Fu, C, & Wang, W. The Effect of Lateral Boundary Treatment of Regional Climate Model on the East Asian Summer Monsoon Rainfall Simulation. Chinese Journal of Atmospheric Sciences (1998)., 5, 779-790 (in Chinese).

[27] Davies, H C, & Turner, R E. Updating Prediction Methods by Dynamical Relaxation: An Examination of the Technique. Quartly Journal of Royal Meteorological Society(1977)., 103, 225-245.

[28] Giorgi, F, Marinucci, M R, Bates, G T & De, Canio G. Development of a Second-generation Regional Climate Model (RegCM2). Part II: Convective Processes and Assimilation of Lateral Boundary Conditions. Monthly Weather Review 1993., 121, 2814-2832.

[29] Wang, X, Zhong, Z, Hu, Y, & Yuan, H. Effect of Lateral Boundary Scheme on the Simulation of Tropical Cyclone Tracks in Regional Climate Model RegCM3. Asia-Pacific Journal of Atmospheric Sciences(2010)., 46, 221-230.

[30] Giorgi, F,Marinucci, M R, & Bates, G T. Development of a Second-generation Regional Climate Model (RegCM2). Part I: Boundary-layer and Radiative Transfer Processes. Monthly Weather Review(1993)., 121, 2794-2813.

[31] Pal, J S, & Coauthors. The ICTP RegCM3 and RegCNET: Regional Climate Modeling for the Developing World. Bulletin of America Meteorological Society(2007)., 88, 1395-1409.

[32] Grell, G A, Dudhia, J, & Stauffe, D R. A Description of the Fifth Generation Penn State/NCAR Mesoscale Model (MM5). NCAR Tech. Note NCAR/TN 398+STR, Boulder, Colorado(1994).

[33] Kiehl, J T, & Coauthors. Description of NCAR Community Climate Model (CCM3). NCAR Tech. Note NCAR/TN-420+STR, Boulder, Colorado, 1996.

[34] Holtslag, A A M, De Bruijin, E I F, & Pan, H L. A High Resolution Air Mass Transformation Model for Short-range Weather Forecasting. Monthly Weather Review(1990)., 118, 1561-1575.

[35] Dickinson, R E, Sellers, A H, & Kennedy, P J. Biosphere-Atmosphere Transfer Scheme (BATS) Version 1 as Coupled to the NCAR Community Climate Model. NCAR Technical Note NCAR/TN-387+STR, Boulder, Colorado, 1993.

[36] Zhong, Z, Hu, Y J, Min, J Z & Xu, H. Experiments on the Spin-up Time for the Seasonal Scale Regional Climate Modeling. ACTA Meteorologica Sinica(2007)., 21, 409-419.

[37] Marbaix, P, Gallée, H, Brasseur, O, & Ypersele J-P. Lateral Boundary Conditions in Regional Climate Models: a Detailed Study of the Relaxation Procedure. Monthly Weather Review(2003)., 131, 461-479.

[38] Wu, G X, Chou, J F, Liu, Y M, & He, J H. Dynamics of the Formation and Variation of Subtropical Anticyclone. Science Press: Beijing, 314pp (in Chinese). , 2002:

[39] Anthes, R A, Hsie, Y, & Kuo, Y H. Description of the Penn State/NCAR Mesoscale Model Version 4 (MM4). NCAR Technical Note, NCAT/TN-282+STR, Boulder, Colorado, 1987.

[40] Giorgi, F, & Marinucci, M R. An Investigation of the Sensitivity of Simulated Precipitation to Model Resolution and Its Implications for Climate Studies. Monthly Weather Review(1996)., 124, 148-166.

[41] Christopher, A D, & Simon, Low-Nam. The NCAR-AFWA Tropical Cyclone Bogussing Scheme, A Report Prepared for the Air Force Weather Agency (AFWA). National Center for Atmospheric Research, Boulder, Colorado, 2001.

[42] Hoskins, B J. On the Existence and Strength of the Summer Subtropical Anticyclones. Bulletin of America Meteorological Society(1996)., 77, 1287–1292.

[43] Liu, Y, Wu, G, Liu, H, & Liu, P. Dynamical Effects of Condensation Heating on the Subtropical Anticyclones in the Eastern Hemisphere. Climate Dynamics(2001)., 17, 327–338.

[44] Seth, A, & Giorgi, F. The Effect of Domain Choice on Summer Precipitation Simulation and Sensitivity in a Regional Climate Model. Journal of Climate(1998)., 11, 2698-2712.

[45] Zhong, Z, Wang, X, Lu, W, & Hu Y. Further Study on the Effect of Buffer Zone Size on Regional Climate Modeling. Climate Dynamics(2010)., 35, 1027 – 1038.

[46] Wang, Y. An Explicit Simulation of Tropical Cyclones with a Triply Nested Movable Mesh Primitive Equations Model-TCM3. Part II: Model Refinements and Sensitivity to Cloud Microphysics Parameterization. Monthly Weather Review 2002., 130, 3022-3036.

[47] Braun, S A, & Tao, W K. Sensitivity of High-resolution Simulations of Hurricane Bob (1991) to Planetary Boundary Layer Parameterizations. Monthly Weather Review(2000)., 128, 3941-3961.

[48] Murata, A, Saito, K, & Ueno, M. The Effects of Precipitation Schemes and Horizontal Resolution on the Major Rainband in Typhoon Flo (1990) Predicted by the MRI Mesoscale Nonhydrostatic Model. Meteorology and Atmospheric Physics (2003)., 82, 55-73.

[49] Hogan, T F, & Pauley, R L. The Impact of Convective Momentum Transport on Tropical Cyclone Track Forecasts Using the Emanuel Cumulus Parameterization. Monthly Weather Review (2007)., 135, 1195-1207.

Chapter 6

Modelling Sea Level Rise from Ice Sheet Melting in a Warming Climate

Diandong Ren, Lance M. Leslie and Mervyn J. Lynch

Additional information is available at the end of the chapter

1. Introduction

Sea level change can arise from fluctuations in ocean basin size and/or water volume. The fluctuations can have many causes, including filling from landslides; water from melting of ice sheets and mountain glaciers; steric water expansion from temperature increases; seabed deformation; and extended dry or wet periods. Here, the focus is on model projections of sea level rise (SLR) from ice sheet melting in a warming climate. The modelling system is SEG-MENT [1,2], which embraces a range of geophysical flows, has a modular design and supports multi-rheology flows. For ice-sheet modelling, the SEGMENT-Ice module incorporates the complexities of both internal flow, and interactions of the ice sheet with its external environment at its upper and lower boundaries, and along its perimeter. Recent applications of SEGMENT-Ice to the Greenland Ice Sheet (GrIS) and the Antarctic Ice Sheet (AIS) show that it simulates well the ice flow patterns in a variety of different spatial configurations, such as slow moving sheet ice, fast moving stream ice, and shelf ice. It also accurately represents many characteristics of the ice sheets, such as internal deformation, basal sliding, ice shelf calving, and temperature profiles within ice.

Quantifying SLR is a major challenge. Two main factors have contributed to the observed global SLR. One factor is the increased melting of land-based ice. The major sources of stored water on land are ice sheets, polar ice caps and glaciers. The other factor is the thermal expansion of the oceans, as warming ocean water expands. A potential third factor, also related to a warming climate, is the greater filling of the sea basin from landslides and soil erosion. The present melting of land ice is comparable with ocean thermal expansion. In a future warming climate, melting is expected to increase. For example, the melting of the GrIS has been identified as a critical, but poorly understood, process in determining global SLR in the 21st century. The Intergovernmental Panel on Climate Change (IPCC) has estimated the GrIS contribution to be ~24 mm by 2100, compared with 1990 levels. This is regarded as an under-estimate, as recent observations indicate that peripheral outlet glaciers are highly sensitive to atmospheric warming. Observational and modelling studies of mass loss from ice sheets, glaciers and the polar ice caps indicate a contribution to SLR of ~0.2 to 0.4 mm/yr averaged over the 20th century. Globally averaged SLR was at a rate of ~1.7 ± 0.3 mm per year over 1950 to 2009 and at a satellite-measured average rate of ~3.3 ± 0.4 mm per year from 1993 to 2009, a marked increase over earlier estimates.

This chapter projects the sea level contribution from ice sheets through to 2100, using SEG-MENT-Ice forced by atmospheric parameters derived from three different climate models (CGCMs). Notably, SLR contributions from AIS, its increased discharge (indicated by in-creased ice flow and calving rate) and another previously overlooked factor (the change of West Antarctic Ice Sheet (WAIS) basin size as WAIS disintegrates), are presented. In 2007, the IPCC AR4 projected that during the 21st century, sea level will rise another 18 to 59 cm, but the IPCC projections include neither "uncertainties in climate-carbon cycle feedbacks nor the full effects of changes in ice sheet flow" [3].

2. Sea level rise from melting of the cryosphere in a warmer climate

The IPCC estimate of 21st Century excluded "future rapid dynamical changes in ice flow." This caveat was added as no ice sheet model had reproduced observations of ice sheet elevation and velocity changes. Calculating the evolution of these changes was not yet possible. Observed rapid changes have several causes. These include penetration of surface melt water to the ice-sheet bed, enhancing acceleration of ice flow [4,5]; sudden disintegration of floating ice shelves with ensuing acceleration of glaciers flowing into the now-removed ice shelf area [6,7]; and penetration of warm water beneath floating ice shelves, causing significant reduc-tions in the buttressing effects of ice on larger outlet glaciers feeding floating ice shelves [8]. Early attempts to model these processes show large possible losses of ice [9,10,2]. The diffi-culties in projecting future sea level change encouraged the glaciology community to under-stand the causes of observed changes, and in a deterministic way such that causal processes are included in numerical predictive ice sheet models. Workshops discussed the required process studies, and the means to improve the skill of existing ice-sheet models [11-13]. It became clear that new field studies of key processes were necessary, along with improved numerical models of ice sheet flow incorporating both a better representation of fast flowing ice, and the processes driving rapid responses of ice sheets. SEGMENT-Ice is one model implementing these improvements and is showing encouraging predictive skill by incorpo-rating new processes such as granular basal sliding (e.g. till rheology in the dynamics of tidewater glaciers), tabular calving and grounding line dynamics. Model results described here are intended to improve upon the present relatively limited value of sea-level projections in the previous IPCC report.

2.1. Regional sea level rise contributions from the Greenland Ice Sheet

This section projects the SLR from the GrIS through to 2100, using SEGMENT-Ice forced by atmospheric parameters derived from three coupled global climate models (CGCMs). The geographical patterns of near-surface ice warming impose a divergent flow field favoring mass loss through enhanced ice flow. The average model mass loss rate during the latter half of the 21st Century is ~0.64±0.06 mm/year eustatic SLR, significantly larger than the IPCC estimate from surface mass balance. The difference is due largely to positive feedbacks from reduced ice viscosity and the basal sliding mechanism present in the ice dynamics model. This inter-model, inter-scenario spread adds ~20% uncertainty to ice model estimates. The SLR is geographically non-uniform and reaches 1.69±0.24 mm/year by 2100 for the northeast coastal region of the United States, amplified by a weakening of the Atlantic meridional overturning current (AMOC). In contrast to previous estimates, which neglected the GrIS fresh water input, both sides of the North Atlantic Gyre are projected to experience SLR. Impacts on major cities on both sides of the Atlantic, and in the Pacific and Southern Oceans, are assessed. The Atlantic Ocean cities are the most affected. Land ice melting likely will increase in a warming climate [14]. The IPCC estimates the SLR contribution from the GrIS will be 24 mm by 2100, compared with 1990 levels. This likely is an underestimate, as during the past decade the large mass loss (reaching ~0.7 mm/yr sea level contribution in 2010) likely is the result of climate warming. The warming has lasted long enough for the already accumulated effects to be irreversible by just several events, such as one or two years of cooler than annual mean temperatures. For example, the 2010 Iceland volcanic eruption did not slow the melting rate. The SEGMENT-Ice GrIS modeling study described here attempts to reduce the uncertainty in quantifying the global SLR from the GrIS, and to explore its regional manifestations. SEGMENT-Ice has a detailed, enhanced treatment of basal and lateral boundary conditions and 'higher order' terms [1]. SEGMENT-Ice was compared with other models in the SeaRISE project, and showed encouraging skill in reproducing and explaining recent, dramatic ice sheet behavior [15,16].

2.1.1. The ice model

SEGMENT-Ice is a component of an integrated scalable and extensible geo-fluid model (SEGMENT). It solves the conservation equations for mass, momentum and energy for the simulation domain, under the rheological relationships of the participating materials. The energy is bundled in a convenient form for considering phase changes and formation of new interfaces (e.g. Griffith cracks and von Mises yielding conditions). In this subsection, the focus is on the GrIS (Fig.1). Parameterization of viscosity is critical for ice creeping. SEGMENT-Ice has two improvements over Glen's ice rheological law, for factoring in flow induced aniso-tropy, and granular basal condition. Flow enhancement by re-fabricating [17] is implemented so that older ice, farther from the Summit, is easier to deform. SEGMENT-Ice also allows a lubricating layer of basal sediments between the ice and bedrock, which enhances ice flow and forms a positive feedback for mass loss in a warming climate [18,19,1]. Because the ocean temperature is higher than around Antarctica, Greenland has no ice shelves. There however are several water-terminating fast glaciers around the peripheral of GrIS, such as Jakobshavn (J), Kangerdlugssuaq (K), Helheim (H), and Petermann (P) glaciers. In SEGMENT-Ice, ocean-ice interactions are parameterized, with freezing point depression by soluble substances, salinity dependence of ocean water thermal properties, and ocean current-dependent sensible heat fluxes included.

As the climate warms, increased air temperature through turbulent sensible heat flux exchange increases surface melting and runoff. Similarly, changes in precipitation affect the upper boundary input to the ice sheet system. For the 200-year period of interest here, major ice temperature fluctuations are near the upper surface of the GrIS. The strain rate, however, can be large near the bottom and/or the surface, so SEGMENT-Ice has a 31 vertical level, stretched grid to better differentiate the bottom and near surface. The uppermost layer is 0.45 m thick near the GrIS Summit, fine enough to simulate the upper surface energy state on a monthly time scale. Because of its location, the GrIS is an important contributor to eustatic SLR, ocean salinity and the North Atlantic thermohaline circulation [20] In SEGMENT-Ice the total mass loss comprises surface mass balance and the dynamic mass balance due to ice flow divergence. Total mass balance is converted to water volume and is used here as a proxy for the eustatic SLR contribution.

In ice flow, inertial and viscous terms counteract pressure gradient forces. The full Navier-Stokes equations are used in the momentum equations of SEGMENT-Ice. Because of compara-rably large aspect ratios, ice streams and surrounding transition zones are the areas where a full Stokes model is needed most [21]. SEGMENT-Ice uses a terrain following coordinate system, the sigma coordinate system, σ (grid lines in inset of Fig. 2), defined as $\sigma = (h - r)/H$, where h is the distance from the ice surface to the Earth's centre and H is the local ice thickness, and r is the independent variable in the radial direction in the spherical coordinate system. A vertical integration of the incompressible continuity equation, with surface mass balance rate and basal melt rate as boundary conditions, gives:

$$\frac{\partial h}{\partial t} = -\frac{1}{R\cos\phi}\int_0^1\left(\frac{\partial u}{\partial\theta} + \frac{\partial v\cos\phi}{\partial\phi}\right)Hds - \frac{u_s}{R\cos\phi}\frac{\partial h}{\partial\theta} - \frac{v_s}{R}\frac{\partial h}{\partial\theta} - w_b + b \qquad (1)$$

where t is time, R is the Earth's radius, θ is longitude, ϕ is latitude, u and v are the horizontal velocity components, and w is the vertical velocity component, which is expanded using the continuity equation, assuming incompressible ice. The subscripts 'b' and 's' mean evaluated at the bottom and upper ice surfaces. Equation (1) diagnoses the temporal evolution of the surface elevation, which also is the ice thickness because bedrock is assumed unchanged over a time scale of several hundred years. The surface elevation varies as a function of velocity fields and boundary sources. The surface mass balance rate, b, also includes basal melt rate. For the AIS, it primarily is the net snowfall (precipitation less sublimation and wind redis-tribution) minus basal melting, in ice thickness equivalent. The change in ice thickness multiplied by grid area is the volume ice loss for that grid. The total ice loss is the summation over the entire simulation domain. In Eq. (1), the first term on the right hand side is the vertical velocity, w, evaluated at the upper ice surface.

2.1.2. Input and verification data

SEGMENT-Ice requires initial conditions and static inputs, such as ice thickness, free surface elevation, and the three-dimensional ice temperature field at the initial time of integration, obtained from the SeaRISE project (http://websrv.cs.umt.edu/isis/index.php), at 5 km horizontal resolution. The bottom geothermal distribution is assumed constant over the 200-year simulation. The SLR contribution from the GrIS is from the total mass balance: i.e. input (e.g., snow precipitation, flow convergence) minus output (e.g., surface melt water runoff, flow divergence to open waters or calving). Atmospheric temperature, precipitation and near surface radiative energy fluxes all are critical factors for the future total GrIS mass balance. The CGCMs provide the meteorological forcing. The large natural variability justifies using extended atmospheric time series to extract first-order feedback GrIS signals in a warming climate. The three independent CGCMs (MPI-ECHAM, NCAR CCSM3 and MIROC3.2-hires (see http://www-pcmdi.llnl.gov/ipcc/about_ipcc.php) are chosen for their relatively fine resolution and for providing all atmospheric parameters required by SEGMENT-Ice. Their projections of precipitation and temperature, two key factors affecting ice-sheet mass balance, produce a large spread in the multi-model assessments [3] by 2100.

In addition to atmospheric parameters, to investigate the ice ocean interactions at the water terminating glaciers (e.g., Jacobhavn, Kangerdlugssuaq, Petermann and Helheim), additional variables such as the ocean flow speed, potential temperature, salinity, and density are needed. Density is dependent as it is a function of temperature, salinity and pressure. For outlet glaciers north of 70°N, CGCM ocean model output at 0, 10, 20, 30, 50, 75 and 100 m depths are interpolated to SEGMENT-ice grids. For the Helheim glacier, which resides in the Sermilik fjord, the 1-km resolution, ice thickness data from SeaRISE indicates a terminus depth of about 700 m, close to the estimate of [22]. Oceanic parameters up to 1000 m are used for this glacier. The University of Hawaii data (http://uhslc.soest.hawaii.edu/jasl.html) are used to evaluate projections of regional SLRs.

2.1.3. Sea level contribution from additional melting of the Greenland Ice Sheet

SEGMENT-Ice is integrated forward in time with climate model meteorological forcing over the GrIS, to provide the total mass loss trend over the 21st Century. Monthly atmospheric forcing and the advanced numerics of SEGMENT-Ice can in principle produce realistic monthly fluctuations in ice sheet properties. However, interannual and decadal climate variations in CGCMs largely are random noise, as are ice model projected quantities on the same time scales. Thus, no attempt is made to compare them with *in situ* observations of the model projections on inter-annual to decadal scales. Monthly SEGMENT-Ice output is averaged to obtain annual mean values. A 21-point binomial smoother is applied to the annual means to remove short-term variability. The smoothed lines in Fig. 2 show the eustatic SLR contribution from the GrIS for the 20th and 21st Centuries. For each model, the atmospheric forcing is under the three non-mitigated IPCC Special Report on Emission Scenarios (SRESs): B1 (low); A1B (medium); and A2 (high rate of emission). Total ice volume is a highly aggregated metric and the trend is the resultant of several factors. As inland GrIS remains cold, the feedback from increased precipitation (see Fig.3b) is significant. Therefore, estimates of the

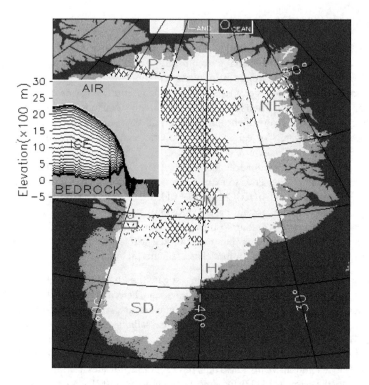

Figure 1. The Greenland Ice Sheet and land cover mask. "L" is land, "O" is ocean. The locations of the Jakobshavn (J.), Kangerdlugssuaq (K.), Petermann (P.) and Helheim (H.) Glaciers, and the North East (NE) ice stream are labelled. Hatched areas are regions with ice loads but with bottom elevations lower than the present sea level. The inset is a zoomed Petermann glacier showing a vertical cross-section.

GrIS contribution to SLR are less sensitive to scenario assumptions before 2030, despite atmospheric forcing diverging after year 2000. After 2030, as atmospheric forcing diverges further, the differences become clear.

The total mass balance terms indicate that the higher precipitation amounts (mm/day) from strong scenario B1 (Fig. 3b) eventually are dominated by increased ice flow divergence and surface melt-water runoff from higher near surface surface temperatures (Fig. 3a). After 2060, the mass loss rate accelerates, as basal sliding becomes significant, especially below the southern tip and the northeast ice stream, signifying a faster mass shed [19]. For the GrIS, the positive feedback from strain heating and reduced ice viscosity may have longer time scales. The consensus is weak scenario B1 has far less total mass loss than the A2 and A1B scenarios, by the late 21st Century.

Sea surface elevation is maintained by atmospheric parameters, and is sensitive to climate changes and it produces corresponding regional sea level adjustments. In addition to factors affecting global SLR, the geographical distribution of SLR adds further complexity, being

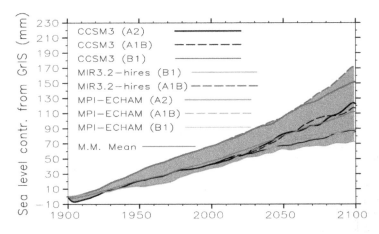

Figure 2. Eustatic SLR (mm) from the GrIS for 20th and 21st centuries, forced by. CCSM3, MPI-ECHAM and MIROC3.2-hires climate models under IPCC scenarios SRES A2, A1B and B1. Red curve is the multi-model multi-scenario mean. Colour shading is model spread.

affected by changes in flow divergence/convergence from ocean currents [23]. For example, owing to the Coriolis force associated with the Gulf Stream, the regional sea surface along the east coast of the United States has a slope tilting seaward. Fresh water discharge from the GrIS weakens the AMOC, suggesting a dynamic adjustment of sea surface elevation. For northeast coastal United States, this produces an additional SLR superimposed on the eustatic SLR. All CGCMs show the largest sea level rebound in the southern Labrador Sea. However, the southern oceans are quite different. With the future strengthening of the Antarctic circumpolar circulation (ACC), the ocean surface slope maintained by the Coriolis force increases, with a significant SLR over an ocean belt at ~46 degree south. The mean change for 2091–2100, relative to 1981–2000, projected by three AR4 climate models under the A1B scenario, reaches 0.4 m over a 10^5 km^2 area of the southern Indian Ocean. A similar, smaller pattern occurs along the east coast of South America. Contributions to SLR from GrIS melting were calculated for 8 coastal cities, calculated from sea level changes with and without GrIS water routing. Southern Hemisphere cities, Cape Town and Sao Paulo, are selected for their proximity to western boundary retroflection currents. Unlike eustatic sea level change, which is approximated empirically [14], quantifying the geographic manifestation of the 0.64 mm/yr global mean SLR requires the inclusion of ocean currents. A CCSM3 sensitivity experiment identified regional GrIS contributions to SLR. The three major contributors: steric, melt water input and ocean dynamics are interconnected; a CGCM is needed to assess the effects of fresh water input. For each scenario, there is monthly forcing from 1900 into the Atlantic Ocean.

The GrIS net mass loss rate matches the time series in Fig. 4 but uses the geographical routing pattern from the ice model. Fig. 4 shows the geographic significance of the SLR, with the sea level time series for three Northern Hemisphere coastal cities: London, New York and San Francisco. Ensemble means of projected sea level change series from CCSM3 after 1900 and

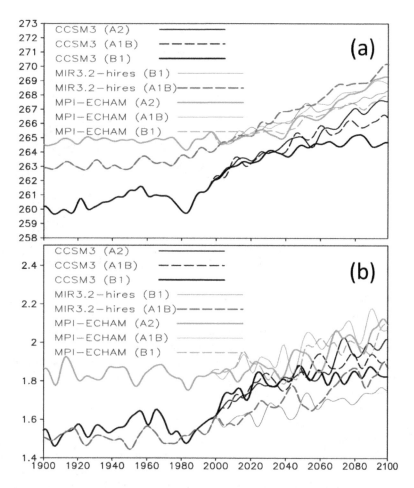

Figure 3. Time series are shown of the 21 point (20 year) low-pass filtered mean near surface temperature (°K, Panel a) and precipitation (mm/day, Panel b) over Greenland, from three climate model simulations with different anthropogenic forcing scenarios. The SRES A2 simulation is not available for the MIR3.2-hires climate model at the time of writing.

expected ensemble mean CGCM projections, neglecting GrIS contributions are compared. The CCSM3 rates are lower than observations (see Fig. 5, which has 8 cities). At 99.5% confidence interval, linear trends for observed and modelled SLRs are: 1.77±0.35 and -0.13±0.7 mm/yr for London; 1.84±0.51 and 0.05±0.1mm/yr for San Francisco; and 2.64±3.05 and 0.31±0.67 mm/yr for New York. GrIS melting reduces the underestimation, especially for Atlantic Ocean cities (1.16±0.7 for New York and 0.47±1.0 mm/yr for London). The greatest rise, for New York City, is from 0.31 to 1.16 mm/yr and is 1/3 closer to reality. For the latter half of this century, the global 0.64 mm/yr SLR increases to ~1.69 mm/year near New York. Fresh water from the GrIS contributes most to SLR northwest of the north Atlantic gyre, with greater impact on cities like New York City. For ocean dynamic adjustment, alone, the London SLR slows then ceases after 2050 as water mass is redistributed to the west coast of the Atlantic Ocean, adapting to a reduced Coriolis force. However, fresh GrIS water causes SLR near London.

In contrast, GrIS melt water is not a major SLR contributor in the Pacific Ocean (e.g., San Francisco, Fig. 5 below) and in the southern oceans (e.g., Sao Paulo) where dynamic sea level adjustments are significant but the GrIS contribution is small.

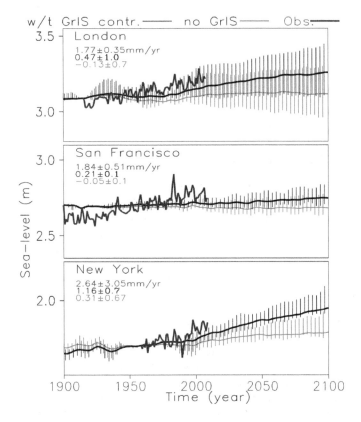

Figure 4. CCSM3 simulated sea level evolution during 1900-2100. Model simulations are performed under the corresponding scenario run but with fresh water routing from the GrIS. The thick, 20-year smoothed black (red) curves are ensemble mean of the multiple model runs with (without) fresh water routing from the GrIS. Vertical bars are the upper and lower envelopes. Blue lines are from tide gauge observations (UHSLC research quality sea level station data). Climate model predictions are shifted so observed and modelled values match at the first observational data grid. The trends and uncertainty range (p=0.05) over the observational period are also given. In generating the fresh water routing scenario, SEGMENT-Ice is forced by CGCMs under different scenarios. For melting fraction of water, the routing scheme of the existing river transport model is used, close to the basin division in [24]. Calving ice is transformed into sea ice fraction with zero salinity.

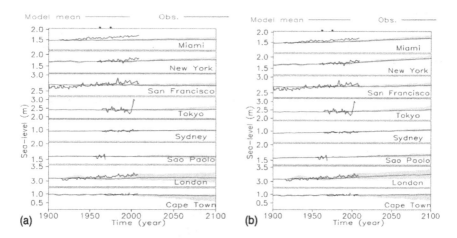

Figure 5. Regional sea level rises near 8 global cities. (a) Without Greenland Ice Sheet fresh water contribution. (b) With Greenland Ice Sheet fresh water discharge

2.2. Eustatic sea level rise contributions from the Antarctic Ice Sheet (AIS)

In this section, the same SEGMENT-Ice system is used as in the previous subsection. However, because of the more complex bedrock-sea water-until configuration, and the increased prevalence of ice shelves, parameterizations that are more physical are activated in the simulations. To minimize the uncertainties associated with the basal granular material, remotely sensed surface velocities are used to deduce the basal mechanical properties and substrata depth [2]. The following is a concise description of the tabular calving scheme and the grounding line dynamics in the SEGMENT-Ice model, and a context is provided for the mass loss and sea level contributions.

2.2.1. An ice shelf approximated by a cantilever beam

An ice sheet is composed of fast-moving, channelized ice streams that drain thick, slower-moving inland ice. Due to the present relatively cold ocean temperatures around the AIS, inland ice streams discharge into the ice shelves, which are the floating extensions of the AIS. The ice shelf mechanics component of SEGMENT-Ice is not available for previous GrIS studies. The ice shelf parameterization in SEGMENT-Ice is derived from an ice shelf's life cycle, which is an advancing-thinning-attrition cycle (Fig. 6). Ice is brittle at higher strain rates, especially under tension, with a melting point diffusivity $\sim 10^{-15}$ m^2/s, much lower than that for elemental metals. For inland ice, crevasses/cracks collocate with locations of concentrated strain rates, for example, after ice flows from relatively gentle into much steeper bedrock slopes (Fig. 6), which is the case for the Pine Island Glacier of the West Antarctic Ice Sheet (WAIS). At the flanks of ice stream, crevasses caused by transverse strain also are prevalent. When ice crosses the grounding line and floats, its contribution to SLR is negligible. Nevertheless, ice shelves are integral components of an ice sheet's thermo-dynamic system. For example, the discharge rate of inland ice is influenced by the buttressing restraint provided by the ice shelves [25].

The following is a flow-dependent ice shelf calving scheme unique to SEGMENT-Ice. Figure 7 illustrates the calving physics implemented in SEGMENT-Ice. The shelf is in near hydrostatic balance with the ocean waters, but the flow structure inside the ice-shelf determines that it is a dynamic scenario of advancing-thinning-breaking, from grounding line toward the calving front. In Fig.6, upper panel (a) is a conceptualization of the Amery Ice Shelf. Compared with land ice, where shear resistance counters most of the surface elevation caused by gravitational driving, ice shelves thin by creep thinning.

Except at the very bottom, the cryostatic pressure inside the ice always is higher than the hydrostatic pressure at the same level from the ocean water. The differences (net horizontal stress) reach a maximum at sea level (the thin red curve in Fig. 6a). Hence, the vertical variation of horizontal shear and the vertical profile of the velocity have a turning point at sea level. The negative vertical strain rate (compression) causes a horizontal divergent (positive) strain rate. Inside the ice shelf, the spreading tendency is restrained mainly by along flow stress. Ice shelves spread under their own weight and the imbalance pushes most at the calving front and around the grounding line. According to [26], this horizontal compression produces an anvil shaped outreach at the calving front [27, 28]. The forward slanting of the Amery Ice Shelf at the calving front (Fig. 6c) is a manifestation of the vertical shear in the ice flow. As shown in Fig.7, although the bulk of the shelf is in hydrostatic balance with the ocean water, the "mushroom" shaped, girder-like spread section usually is not. The limiting length of this portion, before breaking from the main body of the shelf (b_k in the diagram), is limited by the tensile strength of ice (~2 Mpa at present for marginal regions of the Ross Ice Shelf), the ice thickness (H) and the ice creeping speed and vertical shear. In general, the tensile strength of ice can vary over a wide range depending on temperature, strain rate and grain size, as natural ice is polycrystalline. This length is regularly approached, causing systematic calving of an ice shelf. There are random components in the calving processes, such as hydro-fracturing [29], which produce ice shelf lobes. Based on elastic mechanics and a cantilever beam approximation, it is proposed that:

$$b_k = \left[\frac{f \, D_T \, H}{\left(3\rho_i g\left(1 - \frac{C_T \rho_i}{\rho_w}\right)\right)} \right]^{0.5}$$

(2)

where $\rho_w = 1028 \ kg \ m^{-3}$ is the density of sea water, ρ_i is the density of ice, g is gravitational acceleration, C_T is a factor taking tidal and sea wind swelling into consideration, and D_T is a dimensionless factor measuring the ratio of Ice flow shear to surface ice velocity. D_T is a function of ice temperature. The density of ice varies with loading pressure. Ice density is sensitive to overloading because the ice has air bubbles encapsulated during the transition from snow to firn and further into glacial ice. Below, a more rigorous derivation is provided of Eq. (2), based on a cantilever beam approximation. Most AIS peripheral ice shelves are cold, with little surface melting year round, so the tabular calving largely is by gravity cracking. As the shelf falls, the buoyancy increases quickly, so it is rare that a complete cut results from a

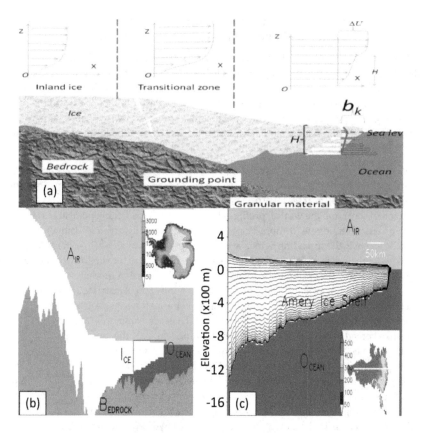

Figure 6. Ice-shelf calving in SEGMENT-Ice. The upper panels in (a) are schematic diagrams of ice profiles, showing the different flow regimes. Note that, due to granular material, the basal velocity is not exactly zero. The acceleration of the ice (to the right) causes the ice to be torn. Inside the ice-shelf, the spreading tendency is restrained mainly by longitudinal stretching (along flow stress). Ice shelves thin by creep thinning. The negative vertical strain rate (compression) causes a horizontal divergent (positive) strain rate. The ice shelf's velocity increases toward the calving front as determined by the spatial integral of the horizontal strain rate. In the diagram, white bulk arrows are stress (hydrostatic pressure) on the right side of the calving front exerted by the ocean, decreasing to zero at sea level. Red bold arrows are static stresses exerted on the left side of the front, decreasing linearly to zero at ice upper/sub-aerial surface. At the shelf bottom, hydrostatic stresses from both sides are almost equal. Red curve is the net horizontal stress, which reaches a maximum at sea level. Hence, vertical variation of horizontal shear and the vertical profile of the velocity have turning points at sea level. The diagram is partially adapted from T. Hughes via R. Bindschadler (personal communication 2011). The vertical profile of the horizontal ice velocity field determines that there will be a "mushroom" shaped spread section not in hydrostatic balance with the ocean water, unlike most of the shelf section. There is a limit to the length of this section before it breaks off from the main body of the shelf (b_k in the diagram). Panel (b) is a cross section of the Amery Ice Shelf (5km DEMs are used in the plotting, indicated in the inset white dashed line). (c) is a zoom-in of the red line confined region in (b), with 1-km resolution ice thickness and surface elevation maps, to illustrate the forward slanting of the calving front. Inset colour shadings are ice thicknesses.

single crack. Periodic forcing from waves and random collision with sea ice and icebergs usually follows fatigue. To generalize these processes and express them as a numerical algorithm, a cantilever beam approximation is used to estimate the calving frequency. As shown in Fig. 7, it is assumed that the material is stiff and the deflection in the z-direction is sufficiently small that linear deformation theory for elastic material is valid. It further is assumed that the cross-section is rectangular to remove variations in the y-direction and work only in the x-z plane (the convention for moment and torque are all in a right-handed coordinate system, as in Fig. 7. The beam material has an elastic modulus E and density ρ.

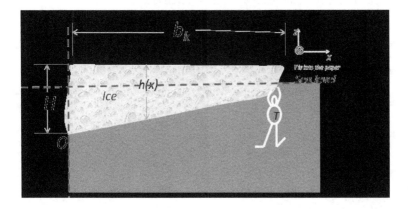

Figure 7. Cantilever beam assumption for an ice shelf. *H* is the ice thickness at the hanging side, which is connected with the main shelf. The *y-z* cross section is assumed rectangular, and ice thickness *h(x)* is assumed linear. *T* denotes external loading, such as tides or random collisions with other icebergs. Small deflections in the z-direction are assumed.

As the cross-section is assumed rectangular, the area moment of inertia is $I = H^3/12$ at O. The momentum around O exerted by the weight of the beam and an external loading (tides and other random factors), T, at the tip end of the shelf is expressed as:

$$M = \int_0^{b_k} x\rho g h(x)dx + Tb_k \qquad (3)$$

where g is gravitational acceleration (9.8m/s²), and M is the moment in the positive y direction. In a static state, the resistance moment should in the *negative* y-direction and of the same magnitude. Because of moment drive, there is potential energy stored around the cross-section passing through O:

$$W = 2EI / Hr \qquad (4)$$

where r is the curvature of the beam at O. At the yielding condition, $f_c = E/r$ is the tensile strength of ice. In Eq. (4), the factor of 2 appears because mass conservation is assumed, so that the cross-sectional area experiencing compression and the area experiencing expansion are the same. The relation between strain and Young's modulus are applied, producing the factor of 2. The stored potential energy and the moment should have the same value. That is:

$$2 f_c I = HM \tag{5}$$

Equation (5) is the key equation for obtaining the limiting length of ice shelf before attrition. For example, if we assume the linear ice thickness profile:

$$h(x) = H - kx \tag{6}$$

where $k = \dfrac{H}{b_k} \dfrac{\rho_i}{\rho_w}$

where ρ_w and ρ_i are respectively the density of water and ice, then substituting into Eqs. (3) and (4), and using Eq. (5), gives:

$$b_k = \frac{-3T}{2a} + \left(\frac{9T^2}{4a^2} + \frac{c}{a} \right)^{0.5} \tag{7}$$

where $a = \rho_i g H \left(1 - \dfrac{\rho_i}{\rho_w}\right)$, and $c = -\dfrac{1}{2} f_c H^2$. For the case without external loading, $b_k = \left[f_c H \Big/ \left(2 \rho_i g \left(1 - \dfrac{\rho_i}{\rho_w}\right)\right)\right]$. This is similar to Eq. (2). In Eq. (2), a more realistic ice profile was assumed in reference to ice flow vertical shear and the further assumption of Eq. (7) was made of assuming T is small compared with the integral part of Eq. (3) (the first term on the right hand side). Ice shelf calving in essentially a fatigue process of visco-plastic ice. In SEGMENT-Ice, the von Mises yielding criteria (a critical point for ice to deform plastically) is applied to identify initial seminal crevasses for inland ice. In principal deviatoric stress form:

$$\sigma_f = \left[\frac{3}{2} \sum_{i=1}^{3} (\sigma_i)^2 \right]^{0.5} \approx 100 Mpa \tag{8}$$

As different parts of the ice-shelf calving front have different ice thicknesses and speeds, it is unlikely a complete cut occurs at once. The typical scenario is that one sector breaks first, and the crack propagates laterally to the neighbouring area. Tides and other random oscillating factors play a role in the speed of the crack, because a crack that does not cyclically open and close does not grow rapidly. Tidal amplitudes around Antarctica generally are small compared with ice thickness except for the portion facing South America. Only those tides with the resonance frequencies of the ice shelves have significant effects on the crack tearing rates. In SEGMENT-Ice, the crack tearing rate is expressed as

$$\frac{da}{dN} = c_1 e^{C_0 |v-v_0|} (\Delta K)^4 \tag{9}$$

Where a is the crack length starting from 0, N is time steps, c_1 is the Paris coefficient [30], C_0 is a negative number indicating the exponential damping of the tidal tearing when it is out of synchronization with the natural frequency of the ice-shelf, v is the tidal frequency and v_0 is resonance frequency of the ice shelf under consideration, and is ΔK the range of stress intensity change (proportional to square of the amplitude of the tide). The resonance frequencies are determined for all peripheral ice shelves.

2.2.2. Grounding line dynamics

In SEGMENT-Ice, ocean-ice interactions are parameterized so that freezing point depression by soluble substances, salinity dependence of ocean water thermal properties, and ocean current-dependent sensible heat fluxes are included. SEGMENT-Ice has a chemical potential sub-model to estimate the effects of ocean water salinity changes on the grounding line retreat of water terminating glaciers and erosion of ice shelves. SEGMENT-Ice uses a molar Gibbs free energy bundle in considering phase changes. Melting/refreezing is determined by the chemical potential of H_2O in both states, across the interface. SEGMENT-Ice estimates ice temperature variations then it calculates the fraction of melted ice. When the terminal heat source becomes a heat sink, freezing occurs and ice extends beyond the initial interface, simulating the grounding line advance/retreat. When newly formed ice is less than the dimension of the grid mesh, it records the fraction that melts first when the heat flux reverses. If newly formed ice fills a grid box, SEGMENT-Ice adjusts its "phase-mask" array, to indicate the new water/ice interface. Melting is analogous.

2.2.3. Mass balance and sea level contribution

The simulated present 3D ice flow field, using the retrieved granular properties, current ice geometry, and current climate conditions (Figures not shown), agrees well with InSAR measurements and provides more information than the InSAR measurements. For example, from the Modified Antarctic Mapping Mission (MAMM, [32]), the Ross Ice Shelf has a large downstream flow speed. The model shows that for most of the flat section of the ice shelf, the ice flow has very small flow-direction accelerations. Only near the edge does the ice flow accelerate toward the ocean; then the vertical velocity profile in Fig. 7 becomes clear. Moreover, in the central thick ice area, there is a stagnant region with ice flow speeds less than 50 m/yr. The small ice velocities for the central area primarily are a result of the relatively large ice thickness. The small dynamic mass balance for the interior of the ice shelf explains why ocean melting mostly balances the precipitation. Consequently, the ice shelf usually is called a buttressing ice shelf, which also prevents direct ocean/ice interaction for inland ice. With the massive rate of basal ablation of the ice shelf by underlying ocean water [33], it is uncertain if ice shelves can maintain near balance in a warming climate.

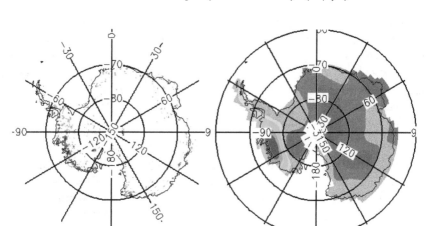

Figure 8. Left panel is the SEGMENT-Ice model simulated surface elevation change between 2003 and 2009 in m/yr (measured as the difference 2009 minus 2003). The right panel is that obtained from the GRACE observations for the same period.

Figure 8 compares the SEGMENT-Ice simulations with the Gravity Recovery and Climate Experiment (GRACE) measurements. The observed mass loss rate for land-based ice in the Antarctica is ~193 km³/year during 2003-2009. Figure 8 (left panel) shows the SEGMENT-Ice simulated geographic distributions of rates of surface elevation changes over Antarctica between 2003 and 2009 (m/yr equivalent water thickness change). Figure 8 (right panel) is the GRACE observations of ice mass change rate over the same period. The post-glacial rebound (PGR) is removed using the method of [31]. A de-correlation filter and a 300-km Gaussian smoothing have been applied to the raw data. Note that the model provides more details than the GRACE observations. That part of West Antarctica facing the Amundsen Sea has a systematic mass loss (>0.6 m/yr in model and the smoothed GRACE observations show >0.1 m/yr reduction in surface elevation) during the five-year period. As a whole, the WAIS is losing mass, but the ridges are gaining very slightly during the five-year observation period, a feature captured by the model, but not well differentiated by the coarse resolution of the GRACE measurements.

To access the credibility of the SEGMENT-Ice mass loss rate estimates of Fig. 8 (left panel), the NCEP/NCAR reanalysis data was merged with atmospheric forcing from CGCMs for 1948-2009. Coupled ocean-atmospheric climate models have difficulty in reproducing the observed interannual and decadal climate variations and cannot be used as climate forcing for

SEGMENT-Ice model validation against observations on those time scales. More realistic climate forcing is provided by the NCEP/NCAR reanalyses [34]. Reanalyses are used by climate researchers as surrogates for real observations on large spatial scales. As mentioned above, Figure 8 (left panel) shows the surface elevation changes of the AIS between 2003 and 2009. Currently, the mass loss in Antarctica is dominated by ice flow acceleration in parts of West Antarctica. In this respect, SEGMENT-Ice simulated mass loss rates are close to GRACE measurements, as well as revealing details that GRACE, limited by its horizontal resolution, cannot identify. For example, the model simulations show clearly it is the peripheral sectors of WAIS that have lost mass most significantly, due to ice dynamics. In addition to changes in surface mass balance, these sectors also have suffered enhanced submarine melting and a consequent grounding line retreat and acceleration. In contrast, the Whitmoor Mountains gain mass from increased snow precipitation. Shelf dynamics likely play a role because peripheral regions that shows significant (>10 cm/yr) mass losses are mostly near major ice shelves: e.g. the Amery and West Ice shelves and the fringing shelves around Dronning Maud Land in East Antarctica. Examining present ice-water geometry near the calving front of the Ross Ice Shelf indicates that, for the next systematic tabular calving to occur, the ice surface elevation must be lower by ~20 m. This indicates that the backward stress it provides to the inland ice is lower than its 'climatological' value. Without considering ocean and atmospheric warming, buttressing gradually will be restored in the next two decades. Despite the very different resolutions, (the model has a 5-km resolution), the overall mass loss rate of ~180 km^3/yr is close to GRACE observations. From Eq. (1), total ice mass loss is comprised of ice flow divergence and boundary mass input/loss (During the reference period (1900-1920) for ice mass loss from climate warming, it is assumed that dynamic mass loss almost equals surface mass balance, with the difference being the interglacial residual trend. With this assumption, variation of modelled total ice mass is a function of anthropogenic climate warming, not past climate. The spread in future scenario, different model runs are also more representative of changes of atmospheric conditions as climate warms, minimizing uncertainty associated with their detailed parameterizations in atmospheric/oceanic physics. Thus, the following best represents the climate warming effect on the AIS total mass balance. Surface elevation changes of AIS between 2000 and 2060 are examined further under the CCSM3 A1B scenario. In general, the peripheral areas are losing ice from climate warming. For the Antarctica Peninsula, a 1 m/yr lowering of the surface elevation is sustainable (Fig. 9); loses more mass because of higher surface air temperatures and summer surface melt. Precipitation is highest but the increase does not compensate for mass loss from extra melting. Because of the higher temperatures, ice flow speeds also are higher. This sector has the most rapid change over the widest area and the greatest impact on total ice mass loss. Some interior regions have significant ice surface elevation increases, e.g. 0.2m/yr upstream of the Ross Ice Shelf. Small-scale elevation increase/decrease pairs in the interior of the ice sheet may persist as they can change sign, likely due to snowfall fluctuations. There are signals from the dynamic flow response to a warming climate. Near the Foundation Ice Stream, there is significant mass gain in the 21st Century from ice flow convergence. The meandering banded patterns of mass loss, most significant near the coast but reaching several hundred kilometres inland, are persistent features of the dynamic response of ice flow to a warming climate.

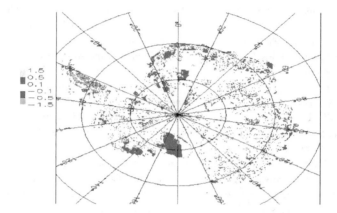

Figure 9. SEGMENT-Ice simulated geographic distributions of rates of surface elevation changes between 2000 and 2060 (m/yr). Peripheral areas and the AP lose mass most significantly. Upper streams of the Foundation Glacier gain mass from increased snowfall. The inland alternative, tributary patterns of mass loss and gain arise from dynamic response to climate warming, corresponding to ice flow convergence/ divergence pattern

The marine based ice sector of West Antarctica, confined by Marie Byrd Land, Siple coast and the Whitmoor Mountains, discharge ice primarily to the Ross Ice Shelf. As the climate warms, increased snowfall partially compensates the effect of flow divergence for the marine based ice. The increases in ocean temperature are ~ 0.2 K for the surrounding oceans, and the increased erosion of the ice shelves from oceans are ~ 1m/yr. The elevation increases caused by net snow accumulation are less than 0.1 m/yr. Horizontal spreading within an ice-shelf is laterally accelerating. This would be a thinning effect for local ice thickness if the shelf has uniform thickness. Interestingly, the thickness profile already has adjusted to this flow pattern and is thinner ocean-ward. Thus, the ice shelf resembles a stream-tube. The thickness change of ice from convergence/divergence is minimal. Consequently, the restoring of the buttressing stress is slowed by oceanic erosion of the shelf thickness, and the discharging of marine-based ice is enhanced. The marine based sector of ice mass loss, which actually is land ice with base elevations below sea level, also has a SLR contribution. There is another mechanism from oceanic warming which causes a reduction in the buttressing effect. Due to the small vertical shear in the flow profile, the calving front is slanted forward (the semi-mushroom shaped structure, right tip-end of Fig. 7c). This means that the ocean-ward section is not fully supported by the buoyancy, and extra weight is placed on the section toward the grounding line, causing this section to be submerged more than is necessary to support its own weight. This trend continues until a critical point is reached and a tabular calving event occurs. At this point, the calved iceberg lowers the weight centre and the remaining ice shelf rebounds upward (less submerged in the water). With this cycle, the buttressing effect fluctuates. Ocean water has a pole-ward temperature gradient. For the fringing ice shelves of Antarctica, the ocean temperature at the calving front is warmer than around the grounding line. If this gradient is larger than $0.0007 \nabla H$ °C/m, where H is the shelf ice thickness, ocean melt also contributes to tabular calving and to a reduced buttressing effect for the inland ice.

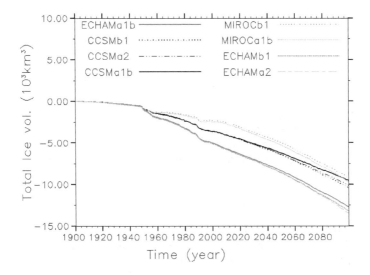

Figure 10. Total ice volume changes obtained by forcing SEGMENT-Ice with atmospheric parameters from three CGCMs, under the IPCC A1B, B1 and A2 scenarios. The 1948-2008 period uses NCEP/NCAR reanalysis data (identical across models and scenarios). Clearly, the inter-model differences are larger than the inter-scenario differences. However, the rates of decrease are similar between the models.

The above results are only for CCSM3 under the SRES A1B scenario. Atmospheric and oceanic forcing parameters were used from two other CGCMs: the MIROC-hires and ECHAM/MPI-AM. These are independently developed models. With 1948-2009 replaced with reanalysis atmospheric forcing, total mass loss rates are different between the CGCMs (Fig. 10). Only the CCSM3 mass loss rates are close to GRACE measurements. The lower MIROC-hires loss rate is due to its warm global air temperature bias of about 2 degrees above observations. Thus, reanalysed temperature field, compared to the MIROC-hires 1900-1920 reference period, is too low and there is less ice melting around the AP, resulting in a slower mass decrease than reality. Systematic biases in ECHAM/MPI-AM temperature and precipitation are lower. Outside the period 1948-2009, mass loss rates are similar among the CGCMs. Using the 1900-1920 reference periods of each climate model, there are large inter-model and inter-scenario similarities in mass loss rates, with a magnitude of spread less than 10% of the absolute decrease, compared to the 1900 level, up to year 2100. Compared with other mountainous regions, the AIS surface is flat, with the steepest slope less than 14 degrees in the 5-km SeaRISE topography data set. This poses less of a challenge for atmospheric components of the CGCMs, and all three models have similar weather patterns over the entire AIS. Region-specific biases thus are not of concern.

The high inter-model and inter-scenario consistency adds confidence in the quantification of future sea level change contributions from the AIS. As ice temperature increases, the viscosity decreases and ice flow increases. Ice flow divergences also increase, resulting in a more significant mass loss. The sensitivity of the AIS ice flow to a warming climate likely results from three positive feedback mechanisms, described by [35]; most notably the positive feedback between granular basal slip and the ice flow. The coastal sectors are increasingly coupled with the interior regions along the preferred channels of ice streams. These ice streams form because of the slow turnover time of the ice that changed its rheological property under gravity on slopped surfaces. The ice along steeper slopes creeps progressively faster, forming an ice stream. Downstream towards the ocean there is granular material formation and accumulation. A warming climate enhances ice deformation and positive feedbacks are triggered among ice flow, granular material accumulation, and reduced resistance to further deformation. These processes provide an explanation for the accelerated rate of mass loss in the 21st century. At the flanks of ice streams, large horizontal transverse stresses create crevasses. The crevasses, when advected downstream to ice shelves, act as seeding cracks that enhance tabular calving.

2.3. Summary

Current CGCMs are not coupled with sophisticated land-ice models, so the uncertainty of the GrIS melting contribution to SLR is large. This study projects the eustatic SLR contribution from GrIS using a new ice dynamics model, SEGMENT-Ice [15]. Forced by CCSM3 atmospheric parameters, the SEGMENT-ice model is integrated for 200 years (1900-2100). The near-surface ice temperature increases for most of the GrIS (Fig. 5, [16]). The greatest warming of over 3°C by 2100, under SRES B1, corresponds to high precipitation areas in a band along the 2000 m GrIS elevation contour. The ice warming decreases inland and reaches a minimum (~0.5°C for SRES B1) at the Summit. As ice viscosity decreases with increasing temperature, the warming pattern adds extra divergence to the original flow field. This ice discharge process probably scales in proportion to surface temperature changes. The average mass loss rate projected by SEGMENT-Ice over the latter half of the 21st century is equivalent to ~0.64mm/ year global mean SLR, which is significantly greater than the IPCC AR4 estimates. The lower limits of the IPCC AR4 estimates (0.01 m, under A1B) therefore should be increased to ~30 mm by 2100, with 95% confidence, assuming other sea level change contributors remain unchanged. To investigate the spatial distribution of the melt water, ice model simulated GrIS mass loss time series are used as input to climate models. The SLR is geographically non-uniform, reaching 1.69 mm/year for the northeast coastal United States, being amplified by a weaker meridional overturning circulation in the Atlantic Ocean. In other oceans, e.g. the Pacific and southern oceans, projected changes are far smaller. Both steric effects and contributions from melting mountain glaciers flatten with warming [36], but the GrIS melting contribution accelerates before declining surface area imposes a limit; this is highly unlikely in the 21st Century. For the AIS, the sensitive and uncertain basal granular material properties are retrieved from InSAR observed surface ice velocities (e.g., using a best fit of model simulated surface velocity and observed surface velocity as retrieval criteria). With improved granular rheological parameterizations, SEGMENT-Ice is driven by atmospheric parameters provided by CGCMs to project of future mass shed from the AIS. There is a high level of inter-model and inter-scenario consistency, with all indicating that the mass loss rate increases with time, and is expected to reach ~220 km³/yr by 2100. Although they have no direct sea level change contribution, ice shelves are integral components of the AIS. Periodic calving, as a normal ablation process, releases tabular icebergs to maintain a dynamic balance of the AIS. In a warming future climate, increased air and ocean temperatures thin the ice shelves by surface erosion, but also they increase the vertical shear near the ice shelf front and cause more frequent tabular calving. Direct erosional ice shelf thinning and ice sheet rebounding, after calving, both signify reduced buttressing effects that lead to further increases in the inland ice mass loss rate. Of note is that when the SLRs from the GrIS and the AIS are combined, the average SLR contribution is +2 mm/yr over the 21st Century.

3. Sea level rise from basin volume change

In Section 2, sea level change from increased input from cryosphere was discussed. Sea levels can change even without an increase of seawater total volume, for example in cases of basin volume changes. There are many causes of basin volume changes, e.g. glacial rebound (PGR), underwater earthquake deformation of the sea bed, and landslides that shed significant amounts of debris into open waters, often also from earthquakes. In the next subsection, a previously overlooked mechanism that may cause faster sea level changes than melting of the cryosphere is discussed. It also is affected by a rapidly changing climate.

3.1. West Antarctica Ice Sheet's possible future disintegration may cause landslides

As it is the largest potential contributor to SLR, quantifying the AIS total mass balance is important to the global hydrological cycle and fragile polar ecosystem consequences. The AIS, especially the West Antarctica Ice Sheet (WAIS), has been extensively studied [37-40]. Much grounded ice in west Antarctica lies on a bed that decreases inland and extends well below sea level (Fig. 11). This bathymetry makes WAIS subject to marine-ice sheet runaway instability [40]. Complete WAIS melting requires $\sim 10^{21}$ J, equivalent to 30,0000 Pinatubo-size volcanoes, which cannot be provided under natural conditions on a century time frame.

Marine-based ice sheets have SLR contributions without melting completely, as they need only be partially afloat, which is possible if the basal melt water is connected to the oceans. At present, the WAIS stability results from the presence of buttressing ice shelves. Because much of the WAIS inland ice has basal melting, the gravitational driving stress cannot be balanced locally. Ice-shelves have flat (upper/sub-aerial) surface elevations and need little resistive stress to maintain balance (except near the tip-end). The hydrostatic pressure from the submerged portion of ice shelves provides the primary resistive stresses for the neighbouring coastal land ice, to balance gravitational driving stress arising from uneven surface topography. Warming from underneath the marine-based ice sheet, particularly that affecting ice-shelf viability could release this potentially fragile stability and lead to accelerated creeping of the WAIS. Ice is brittle at higher strain rates, especially under tension, because its melting point diffusivity is around 10^{-15} m^2/s, which is much lower than the values of $\sim 10^{-11}$ m^2/s for elemental metals. Accelerated creeping thus implies likely breaking up of the WAIS. At the same time, after ice shelves are removed, the pathway for seawater to erode marine based ice sheets becomes open.

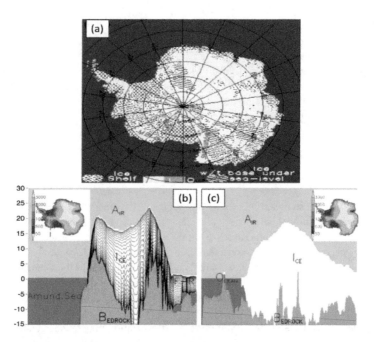

Figure 11. a) Antarctica land-ice-ocean mask based on SeaRISE 5-km resolution digital elevation, ice thickness and bedrock elevation data. Colour shaded white is ice, yellow (brown) is bare ground (L), and blue is ocean. The ice shelves are cross-hatched; land-ice with base under sea level (marine based) is hatched. West Antarctica has more complex ice-water-bedrock configurations than the rest of Antarctica. The WAIS is confined by the Transantarctica Mountains and 40 °W longitude. The Peninsula has a limited ice volume (<3.3x10⁴ Gt) compared with land-based ice of WAIS (~2.8 ×10⁶ Gt). Panels (b) and (c) are the West Antarctica land-ice-ocean mask from SeaRISE 5-km resolution digital elevation, ice thickness and bedrock elevation data. Colour shading is ice (white), bedrock (yellow, brown) and blue (ocean). Panel (b) is a cross-section along the F-R shelf/Amundsen direction (inset; red dashed line). Panel (c) is along the Siple coast direction (inset; red line). In a future warming climate, ocean waters likely enter the WAIS through the Siple coast pathway. The extensive troughs (if ice is removed) can reach depths > 2 km. Colour-shading in insets is surface elevation over the AIS. Marie Byrd Land Ice Cap (MBLIC) and the Whitmoor Mountains ("WM") are labeled.

With the breaking of ice, sectors which have thicknesses below $h_b \rho_w / \rho_i$, where h_b is the bedrock elevation, can float and therefore make an actual contribution to SLR.

Warming factors include drastic increases of geothermal activity from large volcanic eruptions. Although of low probability, they cannot be discarded because of feasibly high impact over a short time period. [42] identified a possible recent active volcano under Ice Stream B (now the Whillans Ice Stream) on the WAIS. Possibly, sub-glacial volcanism could accelerate an existing grounding line retreat, instigating disintegration of the WAIS.

More important is the gradual but widespread oceanic and atmospheric warming driven by anthropogenic greenhouse effects, which is assumed to be salient in the upcoming century (see IPCC AR4, 2007). The likelihood and exact timing of a WAIS collapse are as yet unknown and this study re-examines the SLR resulting from a WAIS collapse, a possibility initially proposed

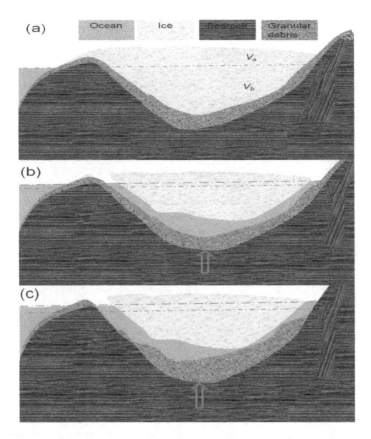

Figure 12. The eustatic sea level rise (SLR) is shown after disintegration of the WAIS. Panel (a) is the ice-bedrock con-figuration before the WAIS disintegration. As the WAIS is disintegrated, the vacant below future sea level will be occu-pied by seawater. The net contribution is the water equivalent of the integrated ice adjusted to account for the volume below sea level (V_b) and for the postglacial rebound of the bedrock (*red arrow, V_r*). Panel (b) is the SLR under this hypothesis.

in [37]. The estimate of [37] is based on the following reasoning, and is illustrated in the sequence of Figs.12a-c, below. As the WAIS disintegrates, the resulting vacancy, which is the space below sea level that was filled by solid ice, then will be occupied by seawater. The net amount of the contribution is the water equivalent of the disintegrated ice, adjusted to account for the volume below sea level (V_b), and for the postglacial rebound of the bedrock (V_r).

Fig. 12b, above, illustrates the SLR under this theory. The SLR can be expressed as $\frac{V_a \rho_i}{\rho_w} + V_b \frac{\rho_i - \rho_w}{\rho_w} + V_r$, where V_a is the ice volume above sea level. This expression slightly relaxes the free-board condition of [37], which assumed that when the summation of the first two terms is positive, the WAIS is floatable. A slight surface elevation decrease, from either negative total surface mass balance or accelerated discharge toward the ice-shelves when oceanic interactions make the ice-shelf thinner, and hence reduce the buttressing effects, will set a significant portion of WAIS afloat. Recently, [39] made a major advance on the original estimate of [37] by using non-static bathymetry from a sophisticated Earth model, and concluded that only a portion of WAIS satisfies the marine ice sheet instability hypothesis, and can become afloat. Inevitably, after the removal of the WAIS, landslides, primarily from Marie Byrd Land, will further change the basin shape, as shown in panel (c). In principle, only the debris situated originally above sea level has a SLR contribution after sliding down-slope to locations below

sea level (V_g). The SLR then becomes $(V_a + V_b)\frac{\rho_i}{\rho_w} - V_b + V_r + V_g$. These studies, however, did not take into account the fact that ice overlaying bedrock is an ideal configuration for rock erosion and the production of a large amount of granular material, especially beneath Marie Byrd Land and the Siple coast). Loading of thick ice above slopes reduces landslide occurrence because of the large confining pressure and because granular debris is effectively cemented by ice crystals. Inevitably, after the removal of the WAIS, landslides will further change the basin shape (Fig. 12c) on time scales much shorter than the basin rebound. In principle, only the debris that is originally above sea level has a SLR contribution when it slides to locations below sea level (R. Alley, Personal Communication 2011). V_g is estimated using the SEGMENT-Landslide model [1], driven by meteorological parameters provided by coupled climate models.

3.2. Data and methods

SEGMENT-Landslide accounts explicitly for soil mechanics, vegetation transpiration and root mechanical reinforcement, and relevant hydrological processes. It considers non-local dynamic balance of the 3D topography, soil thickness profile, basal conditions, and vegetation coverage [15] to project the driving and resistive forces. It describes flow fields and dynamic evolution of thickness profiles of the medium. SEGMENT-Landslide monitoring and predicts landslides and their ecosystem implications. Applications to the polar environment of the WAIS pose fewer challenges for SEGMENT-Landslide because vegetation is not involved. In addition, rainfall morphology is not a concern because solid precipitation dominates. Monthly mean atmospheric forcing parameters suffice for driving the temperature solver, which estimates frozen soil mechanical properties. Three independent CGCMs (MPI-ECHAM, NCAR CCSM3 and MIROC3.2-hires, (http://www-pcmdi.llnl.gov/ipcc/about_ipcc.php) have relatively fine resolution and provide all atmospheric parameters required by SEGMENT-Landslide. Sub-glacial particle properties of WAIS are specified from boreholes and seismic studies (e.g. [43]). The rocks are mostly volcanic and the basalt clasts are of sizes ~10 cm. Loose, ice-cemented volcanic debris also is widespread around Mt Waesche (77°S, 130°W) and the northern Antarctica Peninsula and its constituent blocks. In assigning granular particle sizes, geothermal patterns also are referenced because repeated phase changes at the interface of ice/rock arguably are the most efficient means of erosion and reducing the granular particle sizes. A high-resolution digital elevation map (DEM) is a key input for slope stability analyses.

The SeaRISE project (http://websrv.cs.umt.edu/isis/index.php) provides surface DEM at 5 km horizontal resolution on a South Polar Stereographic projection. The actual sphere resolution is higher at the WAIS, but coarser than 1000 m. Radarsat-1 SAR sensor via MAMM, as a by-product, provides slope information at 200 m resolution. A 200 m resolution surface DEM is obtained by combining SeaRISE and MAMM data. Assuming no major geothermal disturbance from volcanic eruptions, SEGMENT-Landslide simulations indicate that except for very limited northern areas, such as Deception island, where avalanches are likely and involve limited mass redistribution (<10^5 m^3) the present WAIS is stable. In addition to mechanical properties such as particle size, porosity, bulk density, cohesion and repose angle, the granular material thickness is critical in determining the landslide magnitudes. Granular thickness on slopes underneath the present WAIS is inversely retrieved using SEGMENT-Ice [16], constrained by the goodness of fit between model simulated and observed surface ice velocity, over the entire AIS. Figure 13 is the distribution of the estimated granular material thickness over West Antarctica. This is the initial sliding material thickness.

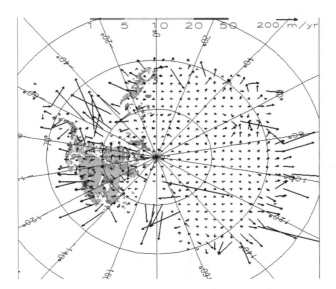

Figure 13. SEGMENT-Ice simulated surface velocities (vector arrows) over the AIS, and retrieved granular material depth under the overlain ice (colour shading, in metres) over the WAIS. The top portion of Mountain Waesche ("W") and the ridge of Whitmoor Mountains ("WM") have shallow debris accumulations (<1.0 m).

Magnitudes of surface ice velocities are sensitive to the thickness of the basal granular material. Based on this sensitivity relationship, repeated projections of the ice model are performed with present ice geometry and ice temperature profiles, but with automatically varying granular layer thickness, to best fit the observed surface ice velocities. Observed ice velocities are obtained from MAMM as described in [32]. As the 7 regional composite of MAMM do not cover the entire AIS (because of the "polar hole"), the metric summation is only over regions with MAMM observations. Upon convergence, the overall agreement between modelled and observed velocities is high, with correlation coefficients of 0.92 for direction and 0.90 for magnitude, respectively. Modelling experiments were carried out to estimate possible landslides contributions to SLR as the WAIS disintegrates. The ice-bedrock configuration of SeaRISE is the same as [45], except they did not consider the contribution of landslides, but all other factors were included. Their estimate is the most recent for the SLR contribution of the WAIS. The SLR contribution is estimated from the following method. SEGMENT-Landslide uses a terrain following coordinate system, the sigma coordinate system. Assuming the sliding material is incompressible, a vertical integration of the incompressible continuity equation, assuming no external sources of granular material, gives:

$$\frac{\partial h}{\partial t} = - \frac{1}{R\cos\phi} \int_0^1 \left(\frac{\partial u}{\partial \theta} + \frac{\partial v\cos\phi}{\partial \phi} \right) H ds - \frac{u_s}{R\cos\phi} \frac{\partial h}{\partial \theta} - \frac{v_s}{R} \frac{\partial h}{\partial \theta} - w_b \qquad (10)$$

Equation (10) is similar to Equation (1) without the surface source term. Equation (10) diagnoses the temporal evolution of the surface elevation and is the sliding material's thickness because bedrock is assumed unchanged over the time scale of several hundred years. It varies because of velocity fields and boundary sources. Changes in surface elevation multiplied by grid area give the mass loss volume for that grid. The total SLR contribution is the sum over all grids with basal elevations above sea level.

3.3. Results and discussion

Figure 14 shows the model-projected surface elevation changes of the bare slopes and, for areas under seawater, the bedrock elevation changes. Landslides can cause some areas to accumulate more than 200 m of sliding material. The most significant regions for volcanic rock and silt debris accumulation are close to Siple coast. However, the source region of the sliding material is primarily the southeast facing slopes of Marie Byrd Land, which contribute ~86% of the accumulated sliding material. Contributions from the Whitmoor Mountains are relatively small (<10%), due primarily to existing granular material on the slopes (Fig. 14). The total volume of the scars on the slopes at elevations above sea level, or equivalently the reduction of the basin volume of the below sea level areas is 3220 km^3, that is ~0.902 cm eustatic SLR. Although small compared with the ~3.3 m eustatic SLR from the collapsed WAIS, it is added to the eustatic sea level with a very short time delay, closely following the ice disintegration. If the fast scenario of [39] is realised, the economic cost to coastally based global cities from the additional 0.902 cm is not a simple linear addition to the 3.3 m SLR.

In contrast, the rebound contribution of the basin bottom takes over 10,000 years, providing a long period for human adaptation and mitigation measures. Landslides considered here primarily are associated with granular material as the ice bonds melt and when the solid ice loading is removed. The two most frequent triggering mechanisms for large-scale landslides, rainstorms and earthquakes, both are negligible. The former is apparent from the year round low air temperatures over the WAIS. Earthquakes are rare in Antarctica, as the Antarctica plate now has only a small portion associated with subduction and is bounded by constructive and conservative margins. Thus, the estimates from this study are expected to be highly representative of the SLR caused by the disintegration of the WAIS. The scarp sizes and the maximum sliding speeds both are sensitive to WAIS disintegration scenarios, but the total volume of the sliding material involved is a conservative property that is insensitive to fast/slow scenarios [39].

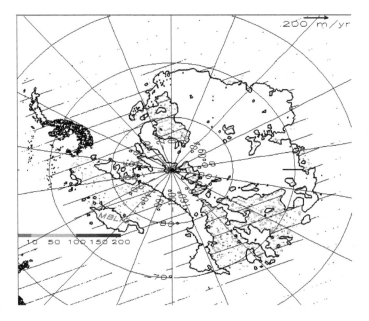

Figure 14. Areas with bedrock elevation increases over 5 m, from landslides accompanying de-glaciation of the WAIS. They are colour shaded (m). The bedrock zero elevation contour lines are shown and areas with bedrock elevations below sea level are hatched. Landslide scars are prominent on the slopes, with localized pairs of elevation increase and decrease. However, regions accumulating most of the sliding material (>5 m depth) are almost below sea level. Marie Byrd Land Ice Cap (MBLIC) and Whitmoor Mountains (WM) are labelled for reference. The southeast facing slopes of MBLIC contribute most of the sliding material.

The model predictions confirm that melting of the WAIS is not possible in the upcoming centuries, but breaking and partial floating of ice is possible, provided the sea water could find pathways to the bottom of ice sectors, which have basal elevations below sea level and low free-board potential. Then the WAIS might disintegrate in a future warming climate. In this modelling study, the potential contribution to eustatic SLR from a collapse of the WAIS is reassessed, and that previous assessments have overlooked the WAIS as a potential major contributor through slope instability if bolstering ice is removed. Overloading ice has a buttressing effect on slope movements in the same way that ice shelves hinder the flow of non-floating coastal ice. Landslide modelling provides estimates of ~9 mm eustatic SLR contribution from WAIS landslides. Note that changes in sea water volumes caused by melting of the cryosphere are persistent but slow processes. Comparatively, the changes of basin sizes caused by landslide filling, from climate warming or from geo-hazards such as tsunamis, are likely to be far quicker, although the SLR contribution likely will be smaller.

4. Coastal line erosion because of sea level change

Shorelines are unique features and are referred to as "Geo-Indicators" in [44]. They never have had long-term positions. Shoreline change has a direct impact on communities, property, industrial and recreational facilities, and species living on both sides of the shoreline. [45] has shown that shoreline change is increasingly affected by climate warming. The relative abundance of glaciers in the current climate suggests that a large SLR is possible from the current cryosphere. As one critical factor contributing to shoreline erosion, the SLR threatens many coastal ecosystems (e.g. [46]). Climate change impacts also include possible increases of sea-surface temperatures and greater variability in patterns of rainfall and runoff [47]. Altered wind patterns and possible changes to wave climate and frequency, storms intensity and duration (see [48]) all accelerate coastal erosion and increase inundation of low-lying areas, and cause saline intrusion into coastal waterways. To predict shoreline changes, simply mapping the bathymetry and coastal surface elevation clearly is inadequate as shoreline susceptibility depends on coastal geological composition, in addition to coastal slope, and are estimated in [49].

The claim that water depth is invariant is appropriate for sandy coasts, has been verified by a number of studies [50]. SLR is accompanied by significant changes in wave heights and tidal ranges. Thus, the relative SLR, wave and tidal changes all should be input to coastal erosion models. This requires that the models in [49] need to incorporate more processes that affect shoreline erosion, because the actual geological features of the coasts are highly complex. Model verifications are now possible because recent advances in satellite imaging and LiDAR remote sensing as described in [51] allow a more cost-effective mapping of shorelines at multiple scales.

5. Conclusions

A natural hazard of increasing concern is a global (eustatic) sea level rise (SLR). Estimating the size and rate of increase of SLR is an urgent observational and scientific challenge. The current cryosphere has the capacity to contribute to the eustatic SLR. Many regions of the Earth, especially low-lying islands and some coastlines, are vulnerable to present SLR. SLR can result from volume variations in ocean basin In this chapter it is shown that a multi-disciplinary modelling approach (Section 1) is needed to investigate the SLR contribution from the two largest ice sheets in the cryosphere: the Greenland Ice Sheet (GrIS) and the Antarctic Ice Sheet (AIS), in a warming 21st Century climate. It was found that the water discharge from the two ice sheets both would increase as the climate warms. For the GrIS, because it discharges water into the northern branch of the Gulf Stream, there are regional manifestations of SLR. The AIS, unlike its northern hemisphere counterpart, has numerous ice shelves that are sensitive to climate change and are of importance for total fresh water discharge into the southern Oceans.

The representation of ice shelves in the SEGMENT-Ice model is described. Using this advanced ice shelf scheme, SEGMENT-Ice simulates well the mass change rate in the past decade using good quality remotely sensed verification data. A projection of the 21st Century mass loss rate is made. From the fresh water discharge alone, the combined contribution from the GrIS and the AIS is equivalent to ~2 mm/yr eustatic SLR in the second half of the 21st Century. The West Antarctic Ice Sheet also has the possibility of disintegrating in a warmer climate. If it does disintegrate, the glacial erosion will produce granular material on the slopes of the bedrock, which may slide into the ocean, causing a reduction in basin volume and a consequent SLR. The magnitude of the SLR from such a bedrock landslide is smaller but is much quicker. Finally, towards the end of the chapter, the problem of coastal erosion was examined, as it is likely to be a common consequence of SLR. The possibility that moderate or severe coastal erosion can have extremely serious socio-economic and natural environmental impacts, it seems prudent that mitigation strategies be developed and implemented as soon as practicable.

Appendix: Cryosphere terms and nomenclature

There is a wide variety of terms applicable to the cryosphere and there is a corresponding nomenclature. The major terms used in this chapter, and their meanings, are presented here in alphabetical order.

Glaciers: Glaciers are large masses of ice that flow very slowly over land. They form from the compacting and recrystalling of snow, when the snow accumulation exceeds combined melting and sublimation. They resemble very slow rivers of ice, crushing rock below, and re-shaping the terrain.

Grounding line: This is the boundary between a floating ice shelf (see below) and the grounded (resting on bedrock) ice that feeds it. When the grounding line retreats inland, water is added to the ocean and the sea level rises.

Iceberg calving: Glaciers flow, or "creep" (see below), under their own weight, and move very slowly like a viscous fluid. When the nose of a glacier reaches a coastline, pieces of the glacier break off ("calve"), producing icebergs that range in size from small to very large.

Ice creeping: Most ice flows, including glaciers, are very slow. Hence, the flows are referred to as ice creeping

Ice Sheet: An ice sheet is a very large mass of glacier ice that covers the surrounding terrain. Commonly, it is defined as having an area greater than 50,000 km^2. There are only two ice sheets at present, the Antarctic and Greenland Ice Sheets. The Antarctic Ice Shelf is by far the larger of the two and contains about 70% of the Earth's entire fresh water supply.

Ice Shelf: An ice shelf is a thick, floating platform of ice that forms where a glacier or an ice sheet flows down to a coastline and then onto the ocean surface. There are numerous ice shelves, the two largest are the Ross Ice Shelf and the Filchner-Ronne Ice Shelf, both located in Antarctica.

Multi-rheology flows: These are flows of various types of fluids. For example, water in its various phases, gas (vapour), liquid (water) and solid (ice) all are present in the cryosphere and treated by SEGMENT-Ice.

Sheet ice: Ice frozen in a relatively thin, smooth, and extensive layer on the surface of a body of water.

Stream ice: Part of an ice sheet that moves faster than the surrounding ice. Stream ice is relatively common in Antarctica and their speed depends partly on the nature of the underlying surface.

Tidewater glaciers: Glaciers that flow into the sea. As the ice moves over the sea it loses support from ground underneath and pieces break off, or calve, forming icebergs. Most tidewater glaciers calve above sea level.

Author details

Diandong Ren[1], Lance M. Leslie[2] and Mervyn J. Lynch[1]

1 Department of Imaging and Applied Physics, Curtin University, Australia

2 School of Meteorology, University of Oklahoma, USA

References

[1] Ren, D, Fu, R, Leslie, L. M, & Dickinson, R. Predicting Storm-triggered Landslides. Bulletin American Meteorological Society (2011)., 92(2), 129-139.

[2] Ren, D, Leslie, L. M, & Lynch, M. (2012). Verification of Model Simulated Mass Balance, Flow Fields and Tabular Calving Events of the Antarctic Ice Sheet against Remotely Sensed Observations. Climate Dynamics 2012; doi:s00382-012-1464-3.

[3] IPCC(2007). Climate Change 2007: The Physical Science Basis. Contribution of Working Group I to the Fourth Assessment Report of the Intergovernmental Panel on Climate Change [Solomon, S., D. Qin, M. Manning, Z. Chen, M. Marquis, K.B. Averyt, M. Tignor and H.L. Miller (eds.)]. Cambridge University Press, Cambridge, United Kingdom and New York, NY, USA, 996 pp.

[4] Zwally, H. J, Abdalati, W, Herring, T, Larson, K, Saba, J, & Steffen, K. Surface Melt-Induced Acceleration of Greenland Ice-Sheet Flow, Science (2002)., 297(5579), 218-222.

[5] Joughin, I, Das, S. B, & King, M. A. Seasonal Speedup Along the Western Flank of the Greenland Ice Sheet. Science (2008)., 320(5877), 781-783.

[6] Scambos, T. A, Bohlander, J. A, Shuman, C. A, & Skvarca, P. Glacier Acceleration and Thinning after Ice Shelf Collapse in the Larsen B Embayment, Antarctica. Geophysical Research Letters (2004). L18402.

[7] Rignot, E, Casassa, G, Gogineni, P, Krabill, W, Rivera, A, & Thomas, R. Accelerated Ice Discharge From the Antarctic Peninsula Following the Collapse of Larsen B Ice Shelf. Geophysical Research Letters (2004). L18401.

[8] Shepherd, A, Wingham, D. J, & Rignot, E. Warm Ocean is Eroding West Antarctic Ice Sheet, Geophysical Research Letters (2004). L23402.

[9] Parizek, B. R, & Alley, R. B. Implications of Increased Greenland Surface Melt Under Global-Warming Scenarios: Ice-Sheet Simulations. Quaternary Science Reviews 20004; 23(9-10) 1013-1027.

[10] Payne, A. J, Holland, P. R, Shepherd, A. P, Rutt, I. C, Jenkins, A, & Joughin, I. Numerical Modeling of Ocean-Ice Interactions Under Pine Island Bay's Ice Shelf, Journal of Geophysical Research-Oceans (2007). C10) 14.

[11] Alley, R, Anandakrishnan, S, Anderson, J, Arthern, R, Bindschadler, R, Blankenship, D, Bromwich, D, Catania, G, Csatho, B, Dalziel, I, Diehl, T, Ferraccioli, F, Holt, J, Ivins, E, Jackson, C, Jenkins, A, Joughin, I, Larter, R, Orsi, A, Parizek, B, Payne, T, Ridley, J, Stone, J, Vaughan, D, & Young, D. (2007). West Antarctic Links to Sea-Level Estimation (WALSE) Workshop. British Consulate-General Houston, DEFRA, Jackson School of Geosciences, British Antarctic Survey, March 2007., 1-3.

[12] Oppenheimer, M, Alley, R. B, Balaji, V, Clarke, G, Delworth, T, Dixon, K, Dupont, T. K, Gnandesikan, A, Hallberg, R, Holland, D, Hulbe, C. L, Jacobs, S, Johnson, J, Leetmaa, A, Levy, H, Lipscomb, W, Little, C, Marshall, S, Parizek, B. R, Payne, T, Schmidt, G, Stouffer, R, Vaughan, D. G, & Winton, M. Report of the Workshop on Ice Sheet Modeling at the NOAA Geophysical Fluid Dynamics Laboratory 8 January 2007. NOAA and Woodrow Wilson School of Public and International Affairs at Princeton University, May (2007). , 1-7.

[13] Lipscomb, W, Bindschadler, R, Bueler, E, Holland, D, Johnson, J, & Price, S. A Community Ice Sheet Model for Sea Level Prediction. EOS Transactions (2009).

[14] Rahmstorf, S. A Semi-Empirical Approach to Projecting Future Sea-Level Rise. Science (2007).

[15] Ren, D, Fu, R, Leslie, L. M, Chen, J, Wilson, C, & Karoly, D. J. The Greenland Ice Sheet Response to Transient Climate Change. Journal of Climate. (2011). , 24(13), 3469-3483.

[16] Ren, D, Fu, R, Leslie, R, Karoly, L. M, Chen, D. J, & Wilson, J. C. A Multirheology Ice Model: Formulation and Application to the Greenland Ice Sheet, Journal of Geophysical Research (2011). D05112, doi:10.1029/2010JD014855

[17] Wang, W, & Warner, R. Modelling of Anisotropic Ice Flow in Law Dome, East Ant-arctica. Annals of Glaciology (1999). , 29, 184-190.

[18] MacAyeal DIrregular Oscillations of the West Antarctic Ice Sheet. Nature (1992). , 359, 29-32.

[19] Alley, R, Dupont, T, Parizek, B, Anandakrishnan, S, Lawson, D, Larson, G, & Even-son, E. Outburst Flooding and Initiation of Ice-Stream Surges in Response to Climatic Cooling: A Hypothesis. Geomorphology (2005).

[20] Alley, R. Ice-Core Evidence of Abrupt Climate Changes. Proceedings National Acad-emy of Science USA (2000). , 97(4), 1331-1334.

[21] Zwinger, T, Greve, R, Gagliardini, O, Shiraiwa, T, & Lyly, M. A Full Stokes Flow Thermo-mechanical Model for Firn and Ice Applied to Gorshkov Crater Glacier Kamchatka, Annals of Glaciology (2007). , 45-29.

[22] Thomas, R, Akins, T, Csatho, B, Fahenstock, M, Goglneni, P, Kim, C, & Sonntag, J. Mass Balance of the Greenland Ice Sheet at High Elevations. Science (2000). , 289-428.

[23] Landerer, F, Jungclaus, J, & Marotzke, J. J. Regional Dynamic and Steric Sea Level Change Response to the IPCC-A1B Scenario. Journal of Physical Oceanography (2007). , 37-296.

[24] Zwally, H, & Giovinetto, M. Balance Mass Flux and Ice Velocity Across the Equilibri-um Line in Drainage Systems of Greenland. Journal of Geophysical Research (2001). , 106, 33717-33728.

[25] Dupont, T, & Alley, R. Assessment of the Importance of Ice-Shelf Buttressing to Ice-sheet Flow. Geophysical Research Letters (2005). L04503.

[26] Reeh, N. On the Calving of Ice from Floating Glaciers and Ice Shelves. Journal of Glaciology (1968). , 7(50), 215-232.

[27] Thomas, R. The Creep of Ice Shelves: Theory. Journal of Glaciology (1973). , 12(64), 45-53.

[28] Hughes, T. Theoretical Calving Rates from Glaciers Along Ice Walls Grounded in Water of Variance Depths. Journal of Glaciology (1992). , 38(128), 282-294.

[29] Scambos, T, Hulbe, C, Fahnestock, M, & Bohlander, J. The Link Between Climate Warming and Break-up of Ice Shelves in the Antarctic Peninsula. Journal of Glaciolo-gy (2006). , 46(154), 516-530.

[30] Timoshenko, S, & Gere, J. Theory of Elastic Stability. 2nd Edition., (1963). McGraw-Hill, New York, USA., 541pp.

[31] Ivins, E, & James, T. Antarctic Glacial Isostatic Adjustment: A New Assessment. Ant-arctic Science (2005). doi:10.1017/S095410200500296

[32] Jezek, K. Observing the Antarctic Ice Sheet Using the RADARSAT-1 Synthetic Aperture Radar. Polar Geography (2003). , 27(3), 197-209.

[33] Bindschadler, R. Hitting the Ice Sheet Where it Hurts. Science (2006). , 311(5768), 1720-1721.

[34] Kalnay, E. and Coauthors ((1996). The NCEP/NCAR 40-Year Reanalysis Project. Bulletin American Meteorological Society 1996; , 77(3), 437-471.

[35] Ren, D, & Leslie, L. M. Three Positive Feedback Mechanisms for Ice Sheet Melting in a Warming Climate. J. Glaciology (2011). , 57(206), 1057-1066.

[36] Raper, S, & Braithwaite, R. Low Sea Level Rise Projections From Mountain Glaciers and Icecaps Under Global Warming. Nature (2006). , 439-311.

[37] Mercer, J. H. West Antarctic Ice Sheet and CO2 Greenhouse Effect. Nature (1978). , 271(5643), 321-325.

[38] Rignot, E, & Bamber, J. van den Broeke M., Davis C., Li Y., van de Berg W., van Meijgaard E. Recent Antarctic Ice Mass Loss From Radar Interferometry and Regional Climate Modelling. Nature Geosciences (2008). , 1(2), 106-110.

[39] Bamber, J. L, Gomez-dans, J, & Griggs, J. A. The Cryosphere (2009). , 3, 101-111.

[40] Joughin, I, & Alley, R. Stability of the West Antarctic Ice Sheet in a Warming World. Nature Geosciences. (2011). , 4(8), 506-513.

[41] Vaughan, D. Recent Trends in Melting Conditions on the Antarctic Peninsula and their Implications for Ice-Sheet Mass Balance and Sea Level. Arctic and Antarctic Alpine Research (2006). , 38(1), 147-152.

[42] Blankenship, D, Bell, R, Hodge, S, Brozena, J, Behrendt, J, & Finn, C. Active Volcanism Beneath the West Antarctica Ice Sheet and Implications for Ice-Sheet Stability. Nature (1993). , 361-526.

[43] Englehardt, H, Humphrey, N, Kamb, B, & Fahnestock, M. Physical Conditions at the Base of a Fast Moving Antarctica Ice Stream. Science (1990). , 248-57.

[44] Lockwood, M. NSDI Shoreline Briefing to the FGDC Coordination Group. January 7, (1997). NOAA/NOS, 29pp.

[45] Walther, G. R, Post, E, & Convey, P. Ecological Responses to Recent Climate Change. Nature (2002). , 46-389.

[46] Meehl, G. A, Washington, W. M, Collins, W. D, Arblaster, J. M, Hu, A, Buja, L. E, Strand, W. G, & Teng, H. T. How Much More Global Warming and Sea Level Rise? Science (2005). , 307(5716), 1769-1772.

[47] Patz, J. A, Campbell-lendrum, D, Holloway, T, & Foley, J. A. Impact of Regional Climate Change on Human Health. Nature (2005). , 438(7066), 310-317.

[48] Knutson, T. R. and Coauthors. Tropical Cyclones and Climate Change. Nature Geosciences (2010). , 3-157.

[49] Bruun, P. Sea Level Rise as a Cause of Shore Erosion. Journal of Waterways and Harbors Division (1962). , 88-117.

[50] Dean, R. G, & Maurmeyer, E. M. Models of Beach Profile Response. In: CRC Handbook of Coastal Processes [Komar, P.D. (ed.)]. CRC Press, (1983). Boca Raton, FL.

[51] Sampath, A, & Shan, J. Building Boundary Tracing and Regularization from Airborne LiDAR Point Clouds. Photogrammetric Engineering and Remote Sensing (2007). , 73(7), 805-812.

Paleoclimate and Geo-Environments

Fractal Nature of the Band-Thickness in the Archean Banded Iron Formation in the Yellowknife Greenstone Belt, Northwest Territories, Canada

Nagayoshi Katsuta, Ichiko Shimizu, Masao Takano,
Shin-ichi Kawakami, Herwart Helmstaedt and
Mineo Kumazawa

Additional information is available at the end of the chapter

1. Introduction

Banded iron formations (BIFs) are chemically precipitated deposits on the Precambrian sea floor and are characterised by alternations of repeat Fe-rich and Si-rich layers [1]. Temporal variations in the volumes of BIFs are considered to be related to early evolution of the atmosphere, oceans, life and the Earth's interior [2,3]. In general, BIFs contain various scales of banding. Bands with a thickness of several tens of meters to meters, a thickness of centimetres and a thickness of submillimetre to millimetres are named macrobands, mesobands and microbands, respectively [4]. Some depositions are related to periodic phenomena, such as annual cycles [4], tidal and solar cycles [5–7], and Milankovitch cycles [8, 9] in the Precambrian. On the other hand, quantitative analysis of the banding is limited to Paleoproterozoic Hamersley (Superior-type) BIFs, although BIFs occur within an age range from 3.8 Ga to about 0.7 Ga [10]. Therefore, it is necessary to investigate different BIFs, in terms of both their age and type, clarified by size and lithological facies (i.e., Superior- and Algoma-types) to understand the nature of their banded structures.

In this study, we analysed the banded structures in Archean BIFs using a nondestructive micro-X-ray fluorescence (XRF) imaging technique. This technique has been used recently to determine the distribution of major and trace elements in Quaternary sediments and Phanerozoic sedimentary rocks for characterising the paleoclimatic and paleoenvironmental signals [e.g. 11-15]. It has also been used in the analysis of BIF bandings [16–19]. Sakai *et al.* [16] reported on an XRF imaging analysis conducted on an Antarctic BIF that showed a clear striped structure of alternating Fe-rich and Si-rich layers. Matsunaga *et al.* [17] investigated the influence of alternation and weathering on the banding in terms of the elemental distribution and the chemical forms, showing that the titanium in the BIF preserved the primary depositional structures and that the chemical states of the iron and manganese present alternated with rhythmic changes in the banding. Fukuda *et al.* [18] revealed that titanium-rich layers in the Hamersley BIF are composed of Ti in silicate and titanium oxide layers and suggested that the oxide phase had precipitated in solution, as well as that the silicate phase had originated from a clastic input of terrigenous origin. Pufahl and Fralik [19] found millimetre-scale chemical grading of Fe, Si, Mn, and Al in the mud lamina of a BIF deposited in Paleoproterozoic shallow water. They suggested that these chemical structures resulted from changes in the Fe^{2+}/Mn^{2+} ratio in the water column, along with changes in the concentration of dissolved O_2, together with the precipitation of inorganic SiO_2 and a rainfall input of terrigenous clay. These previous studies on BIFs using XRF imaging techniques focused on internal structures having mesoband units. However, we are interested in a wider range of patterns in the fluctuations of BIF laminations [20].

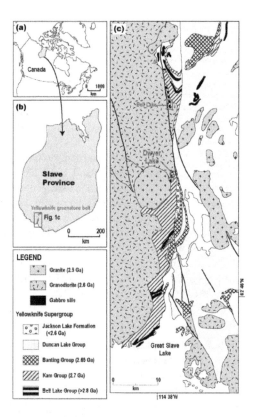

Figure 1. A simplified geological map of the Yellowknife greenstone belt located in the southeastern part of Slave Province, Northwest Territories, Canada, modified from MacLachlan and Helmstaedt [21]. The U–Pb ages are after Isachsen and Bowring [22].

The purpose of this study was to investigate the abundance and variations of the constituent materials in the BIFs, both in the mesobands and on longer spatial scales. In this study, we used an Algoma-type Archean BIF sample with a thickness of about 1.14m, which was collected from a >2.8 Ga Bell Lake Group exposed in the Dwyer Lake area of the Yellowknife greenstone belt, Northwest Territories, Canada (Fig. 1). Based on the results of our XRF imaging analysis and from petrographic observations, we have evaluated whether the banding is a primary or secondary structure, as the constituent materials of the BIF had been completely recrystallized into medium-grade metamorphic minerals under amphibole-facies conditions. After measuring the band-thicknesses of the Fe-rich and Si-rich mesobands using XRF imaging, we investigated the implications of the data for the morphology of the bandings recorded in our BIF samples.

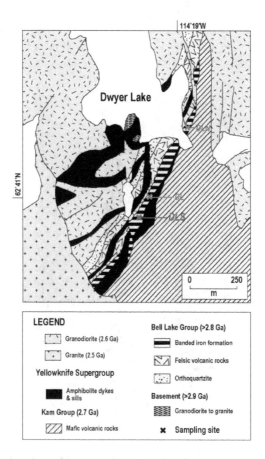

Figure 2. A simplified geological map of the Dwyer Lake area, modified after MacLachlan and Helmstaedt [21]. The U–Pb ages are after Isachsen and Bowring [22]. The sampling points are denoted by the sample names: *DLN*, *DL*, and *DLS*.

2. Samples and methods

2.1. Sampling sites and sample collection

The Yellowknife greenstone belt is the southernmost of approximately 26 granite-greenstone belts in the Slave craton (Fig. 1), and is approximately 35km long and 8–10km wide. The belt is 12km thick, steeply dipping southeast and younging, with homoclinical sequences of calc-alkaline to tholeiitic metavolcanic rocks and metasedimentary rocks [21]. The main volcanic suite of the Yellowknife greenstone belt, composed mainly of tholeiitic pillow lavas, is the Kam Group (ca. 2.7 Ga). The metasedimentary rocks below the granodiorite basement and above the Kam Group (Fig. 2) are divided into the Bell Lake group, which was informally named by Isachsen and Bowring [22]. Here, the term Bell Lake "Group" is used to indicate clearly that it has the rank of a group [20]. The Bell Lake Group is the lowermost part of the Yellowknife Supergroup and comprises orthoquartzite, felsic volcanic rocks and BIF packages from the bottom to the top. In general, the Bell Lake Group is considered to have been formed in a depositional setting, similar to that of a back-arc basin adjacent to a continental margin [21] during transgression [23].

Figure 3. A field photograph of the Dwyer Lake BIF. The light and dark bands on the tabular samples correspond to the Si-rich and Fe-rich mesobands, respectively. The numbers on the sample's surface denote the distance described in intervals of 5cm.

The Bell Lake Group is well-exposed in the Bell Lake and Dwyer Lake areas (Fig. 1). The ages of these regions were determined using U–Pb zircon dates [24]. The felsic volcanic rock below the BIF in the Dwyer Lake region has been dated at 2853 +2/–1 Ma. In the Bell Lake region, the 3m-thick layer of felsic tuff within the 40m-thick BIF unit has an estimated age of 2826 ± 1.5 Ma. In these two regions, we collected continuous BIF tabular samples from the steeply dipped outcrops using a portable diamond cutter (Fig. 3) [14]. In the Bell Lake region, samples from five sequences were taken from the island inside the lake ([20], Fig. 1). On the other hand, in the Dwyer Lake region, the BIF samples were acquired from three sites (*DLN, DL,* and *DLS*) that were in close contact with the pillowed volcanic rocks (Fig. 2). The sample taken from *DLN* (62°41'174''N, 114°18'885''W) was approximately 112cm long, the sample taken from *DL* (62°40'609''N, 114°18'740''W) was approximately 113cm long and the sample taken from *DLS* (62°40'547''N, 114°18'874''W) was approximately 114cm long. In the laboratory, the acquired BIF samples were shaped into plates with an area of approximately 20×20 cm^2, and a thickness of 3cm for XRF imaging analysis [14]. Hereafter, the sample from the Bell Lake and Dwyer Lake regions will be referred to as the Bell Lake and Dwyer Lake BIF, respectively. In this paper, we present a sequential profile of the *DLS* sample, because the Dwyer Lake BIF samples commonly have a similar pattern of its laminations.

2.2. Analysis

The constituent minerals were identified using an optical microscope, a TOPCON DS-130C scanning electron microscope equipped with a HORIBA EMAX-770X energy dispersive X-ray spectrometer at Nagoya University (Japan) and a microfocused X-ray diffractometer (XRD) (Rigaku PSPC/MDG CN2175-A1) at Nagoya University, equipped with a CrK radiation source (wavelength = 2.2909 Å) and a Cr anode target. The diffraction patterns were matched with data from the Joint Committee on Power Diffraction Standards database [25]. The chemical composition of the amphiboles was analysed using a JEOL JCMA-733 electron-probe micro-analyser at Hokkaido University (Japan), operating at an accelerating voltage of 15kV and a beam current of 20nA. The X-ray intensities were converted to chemical abundances according to the oxide ZAF correction. The concentration of Fe^{2+} and Fe^{3+} was calculated from the total FeO content [26].

We used a HORIBA XGT-2000V scanning X-ray analytical microscope (SXAM) [27] at Nagoya University to extract information from the lamination patterns on the BIF sample surfaces. The high-intensity incident X-rays were emitted from an Rh anode at 50kV and 1mA, focused into a microbeam using an X-ray guide tube with a diameter of 100μm, and irradiated perpendicular to the sample surface. The XRF from the sample surface was analysed using the hi-Si detector of the energy-dispersive spectrometer, and the transmitted X-rays were measured using a NaI scintillation detector. In the SXAM system, the sample was mounted on a motor-driven X–Y stage and placed in an open space outside the vacuum chamber, so that there was little limitation on the sample size used. A surface area up to 200×200mm^2 in size was scanned by the incident X-ray beam, and the XRF intensity data were stored at a resolution of either 256×256 pixels or 512×512 pixels. In the 512×512 pixel mode, seven elements between Na and U, as well as the transmitted X-rays, could be analysed simultaneously. Alternatively, 31 elements, as well as the transmitted X-rays, could be analysed in the 256 × 256 pixel mode. The standard measurement time required to acquire a 512 × 512 pixel image was about 40 hours.

3. Results

3.1. Distribution of the major elements

The major elements in the Dwyer Lake BIF samples detected by our SXAM analysis were Fe, Si, Ca, Mn, K, P, S, and Ti (Fig. 4a). Because the concentration of Ti was relatively low, the first seven elements were selected for SXAM imaging analysis in the 512 × 512 pixel mode. The weight percentage of these seven elements was estimated to be > 95%. The scanning area for the imaging analysis was fixed at 51.2 × 51.2mm², and the scan step size was 100μm. The XRF images were acquired allowing overlapping with the adjacent sections above and below to obtain continuous sequence data. The XRF images were reduced to one-dimensional profiles using lamination trace techniques [28] that enabled us to obtain the average value along the deformed bedding planes. After the weighted averages of the overlap sections had been calculated, the sequential profiles shown in Fig. 5a were obtained.

In the XRF images shown in Fig. 4b, the Fe-rich and Si-rich mesobands correspond to the green and white layers shown in the outcrop in Fig. 3, respectively. Fe, Ca, and Mn were enriched in the Fe-rich mesobands, while the Si-rich mesobands contained thin Fe-rich layers (millimetre bands) 0.3–2.0mm thick. Potassium was concentrated locally in the strikingly deformed and altered layers, while phosphorus occurred in thin layers in the Fe-rich mesobands together with Ca. Sulphur was irregularly distributed in both the Fe-rich and Si-rich mesobands. Because K, P, and S occurred in low concentrations and showed no distinct patterns, they are not shown in the XRF images in Fig. 4b.

3.2. Petrology and mineralogy

The Fe-rich mesobands were composed of actinolite, magnetite, quartz, and trace amounts of apatite and pyrite (Fig. 6a). The Si-rich mesobands were composed of quartz, magnetite and trace actinolite. The nomenclature used for the amphiboles follows Leake *et al.* [29]. The Fe-rich mesobands corresponded to the layers shown in the XRF maps of Fe, Ca, and Mn.

Magnetite grains in the Fe-rich mesobands formed several hundred μm-thick layers (Fig. 6a). The Si-rich mesobands were composed of quartz and magnetite, with minor actinolite (Fig. 6b), which correspond to the layers shown in the Si map. The constituent minerals in the

Dwyer Lake BIF samples correspond to the medium- to high-grade metamorphic zones described by Klein [10]. Magnetite grains in the Si-rich mesobands were either included in the quartz matrix or arranged as microbands (Fig. 6b).

The actinolite in the Fe-rich mesobands was green in colour and was either euhedral or subhedral (Fig. 6a). The dimensions of the actinolite were typically approximately 100μm in the direction of the crystallographic *c*-axis and <50μm in the direction perpendicular to the *c*-

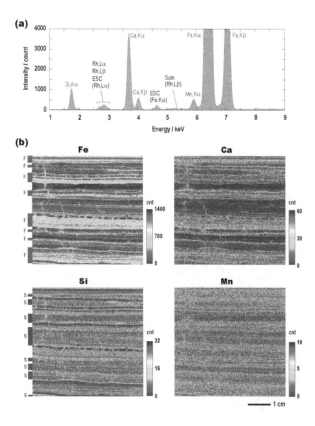

Figure 4. Micro-XRF analysis of a BIF sample (*DLS* 35–45). (a) Cumulative energy spectrum captured using SXAM for an Fe-rich mesoband. The terms ESC and SUM denote the escape and the sum peak, respectively. The Rh peaks denote signals from the Rh target. (b) XRF images of a BIF sample. Key: F = Fe-rich mesoband, and S = Si-rich mesoband. The size of the digitised images represents 512 × 512 pixels. The step size in the *x–y* scan is 100μm. The XRF intensity is represented by the X-ray photon count (cnt).

axis. The actinolite lacked any exsolution textures. Actinolite also occurred locally in the quartz matrix of the Si-rich mesobands. The magnetite grains in Fe-rich mesobands and in microbands within the Si-rich mesobands were relatively large (length = approximately 100μm), and were mostly oblate. Outside the Fe-rich mesobands and microbands, magnetite occurred in small grains included within polygonal quartz grains or at the boundaries of quartz grains. The quartz showed an undulatory extinction.

The schistosity was determined from the preferred orientation of the actinolite grains in the Fe-rich mesobands and in the quartz matrix, and by the preferred orientation of the oblate magnetite. The schistosity was subparallel to the banded structures in the Dwyer Lake BIF samples.

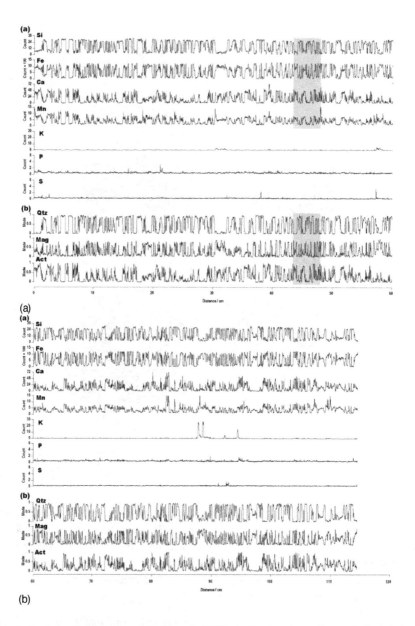

Figure 5. Chemical profiles of the Dwyer Lake BIF samples (*DLS*) in the direction orientated perpendicular to the banded structures. (a) Elemental profiles. The relative abundance of Si, Fe, Ca, Mn, K, P and S are represented by the X-ray photon counts acquired during the micro-XRF analysis. (b) Mineral profiles are calculated from the XRF intensity for Si, Fe, Ca, and Mn. The shaded area corresponds to the XRF images in Fig. 4 and the mineral mode profiles shown in Fig. 7. The abbreviations of the mineral phases follow Kretz [35].

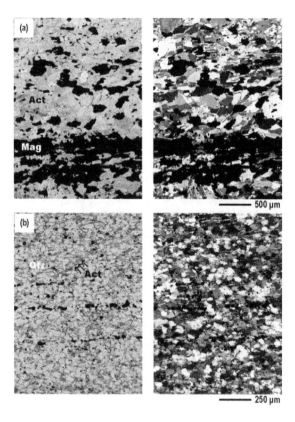

Figure 6. Photomicrographs of BIF sections (*DLN*) cut normally to the banded structures. Left: plane-polarised light. Right: crossed-polarised light. (a) Fe-rich mesoband and (b) Si-rich mesoband.

3.3. Modal compositions

The measured XRF intensity in the element images for a given pixel is related to the bulk concentration of the elements of the constituent mineral for that pixel. Therefore, we assumed that the XRF intensity was linearly proportional to the volume proportion in the mineral, and therefore determined the modal proportions of the constituent minerals in the Dwyer Lake BIF samples (Figs 5b and 7) using the algorithm of Togami *et al.* [30]. The minor minerals were ignored in our calculations and we approximated the samples as being a mixture of three minerals: quartz, magnetite and actinolite. The modal proportions of these three minerals were determined from the XRF intensity profiles of Si, Fe, Ca, and Mn (Fig. 5a). The minor phases of apatite and pyrite were estimated to be <1%. Details of the computation methods are described in Katsuta *et al.* [20].

The modal proportion of actinolite was typically 0.6–1.0 in the Fe-rich mesobands, and <0.2 in the Si-rich mesobands. The modal proportion of magnetite was 0.3–0.8 in the Fe-rich mesobands and approximately 0.7 in the Si-rich mesobands. The magnetite peaks shown in Fig. 7 correspond to a high Fe content in the XRF map (Fig. 4b). The distributions of Ca and Mn in the XRF maps show high concentrations of actinolite. The relationship between actinolite and magnetite in the Fe-rich mesobands and that between magnetite and quartz in the Si-rich mesobands show an inverse correlation.

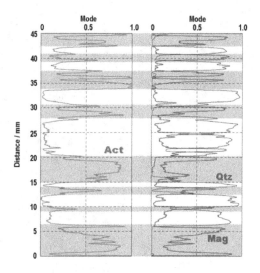

Figure 7. Mineral mode profiles corresponding to the XRF images shown in Fig. 5. The shaded and white bands indicate Fe-rich and Si-rich mesobands, respectively.

3.4. Band-thicknesses

The thickness of the Fe-rich and Si-rich mesobands was analysed using the element and mineral profiles (Figs 5 and 7). The results of our analysis are shown in Fig. 8. The Fe-rich mesobands had a thickness range of 0.1 to 1.4cm (average = 0.35cm) and the Si-rich mesobands had a thickness range of 0.1 to 1.1cm (average = 0.25cm). A thickness of approximately 0.1cm was the cutoff scale for these two mesobands. The number of these mesobands that could be identified was 190 over the entire *DLS* sequence (114cm in length). The frequency distribution showed an exponential decrease in the number of thicker bands.

4. Discussion

4.1. Influence of metamorphism on banded structures

The Dwyer Lake BIF suffered amphibole-facies metamorphism and the constituent minerals were completely recrystallized into medium-grade metamorphic minerals (Fig. 6). The

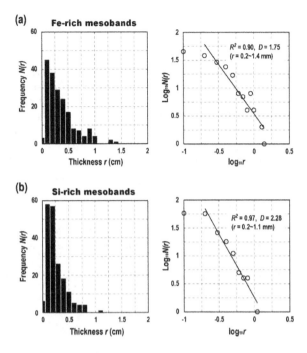

Figure 8. Frequency distributions of the band-thicknesses of the Dwyer Lake BIF samples (*DLS*). Left: a linear scale. Right: a logarithmic scale. (a) Fe-rich mesobands. (b) Si-rich mesobands. *R*: correlation coefficient. *D*: fractal dimension.

schistosity defined by the preferred orientation of elongated actinolite and magnetite was approximately parallel to the boundary between the Fe-rich and Si-rich mesobands. Because of this, it was difficult to evaluate whether the bandings were primary sedimentary structures or secondary metamorphic layers.

In a previous study, we investigated the banded structures in Bell Lake BIF samples [20], which have the same stratigraphic position (Bell Lake Group) as the Dwyer Lake BIF samples (Fig. 1). The Bell Lake BIF had metamorphosed to amphibolites facies and had the schistosity subparallel to the banded structures, as the Dwyer Lake BIF had. Meanwhile, in the Bell Lake BIF samples, we found that the schistosity clearly intersected the boundary between the Fe-rich and Si-rich mesobands in a locally developed intrafolial fold. This implies that the alternating structures of the Fe-rich and Si-rich mesobands already existed at the time of peak metamorphism.

In the Bell Lake BIF samples, the Fe-rich mesobands were mainly composed of hornblende, grunerite, and magnetite. The hornblende was concentrated in the middle part of the Fe-rich mesobands, and was sandwiched between grunerite layers at the margins of the Fe-rich mesoband, which suggests that metamorphic differentiation occurred in the Fe-rich mesobands. The mineral phases and petrographic textures are markedly different from those in the Dwyer Lake BIF samples.

The differences between the Bell Lake and Dwyer Lake BIF samples could have arisen from different initial bulk compositions and perhaps from different metamorphic temperatures. It has been suggested that the Bell Lake BIF had possibly been subjected to a stronger influence of metamorphism compared with the Dwyer Lake BIF. Therefore, we regard the banding and band-thickness of the Dwyer Lake BIF to be a primary sedimentary structure. As has been discussed for the Bell Lake BIF [20], the source of the constituent materials in the Fe-rich mesobands was interpreted as being from mafic pyroclastic materials arising from submarine volcanic eruptions, given that the Bell Lake Group was formed at a time of continental breakup and rifting.

4.2. Causes of the banding in the Dwyer Lake BIF

As shown in Fig. 8, the frequency distribution of the band-thickness in the Fe-rich and Si-rich mesobands in the Dwyer Lake BIF shows an exponential decrease in the number of thicker bands. This suggests that the band-thickness follows a power law distribution. Hence, we determined the scaling exponent, D, of the band-thickness, r, (above the 0.1cm cutoff scale) using the equation

$$N(r) = a \cdot r^{-D}, \tag{1}$$

where N is the number of bands and a is a constant. Consequently, the value of D for the Fe-rich mesobands was 1.75 with a correlation coefficient of $R = -0.95$ and the value of D for the Si-rich mesobands was 2.28 where $R = -0.98$. This statistically significant result implies that the Fe-rich and Si-rich mesobands have a fractal nature.

It is well known that the BIFs have hierarchical structures comprising macrobands (meters thick), mesobands (centimetres thick), and microbands (millimetres thick) [4]. On the other hand, evidence that BIFs have fractal dimensions of the band-thickness, i.e., D, has also been observed in the Archean BIF, Kola Peninsula, Russia [31]. Goryainov et al. [31] reported that the magnetite bands had a thickness distribution of $D = 1.74 \pm 0.1$. This is comparable to that of Fe-rich mesobands in the Dwyer Lake BIF ($D = 1.75$).

In the Dwyer Lake BIF, the most striking feature is that not only the Fe-rich mesobands but also the paired Si-rich mesobands exhibit a fractal nature. In general, it is believed that the absence of silica-secreting organisms in the Precambrian seawater may have led to it be saturated with respect to dissolved silica and the silica was maintained at a saturated level throughout periods of both high and low iron supply [32]. If this model is applied to the bandings in the Dwyer Lake BIF, then the thickness and spacing of the Fe-rich mesobands could be interpreted as indicating deposition by a nonstochastic process.

To our knowledge, the geological evidence that bed thickness follows a power law is limited to observations on turbidite sequences [33] and biocalcarenitic tidal dune successions [34].

However, such depositional processes are different from those of the Fe-rich mesobands in the Dwyer Lake BIF. This is because the constituent materials of the successions are clastic sediments that were supplied from a continental shelf by turbidity currents and deposited in a shelf embayment dominated by tidal currents. Accordingly, we must assign another cause for the morphology of the bandings preserved in the Dwyer Lake BIF.

Presently, we consider the hypothesis of Morris [32] to be the most suitable depositional model for explaining the fractal nature of the Dwyer Lake BIF. It is well known that the banding of mesoband types is well developed in the Hamersley BIF of Western Australia. According to Morris [32], the Fe-rich mesobands resulted from the periodic convection-driven upwelling of pyroclastic materials from a mid-ocean ridge (MOR) or hot spot. Moreover, Morris concluded that the different mesoband types that are intermediate in scale between mesobanding and microbanding were produced by a modified deep-water supply because of varied MOR activity or partial blocking of upwelling water. In the Dwyer Lake BIF, we consider that such an irregular supply of deep seawater may have resulted in the formation of the fractal-like bandings. In addition, the difference in the bandings between the Dwyer Lake and Bell Lake BIFs that have a fractal-like and hierarchical structure [20], respectively, could be explained by the hypothesis of Morris [32].

5. Conclusions

We have investigated the bandings in Archean metamorphic BIF samples collected from the Dwyer Lake area in the Yellowknife greenstone belt, Canada. Chemical and petrographic analysis revealed the following.

1. The Fe-rich and Si-rich mesobands in the Dwyer Lake BIF are regarded as being primary structures formed before metamorphic alternation, based on a comparison with Bell Lake BIF samples.

2. The band-thickness of both the Fe-rich and Si-rich mesobands showed a power law distribution, suggesting a fractal-like nature.

3. The thickness and spacing of the Fe-rich mesobands could have been created by an irregular upwelling of submarine pyroclastic materials.

Acknowledgements

We express our cordial thanks to John A. Brophy and Bill Padgham, Northwest Territories Geoscience Office, for their support in our research activities in Canada. We thank S. Hori, T. Okaniwa, H. Yoshioka, A. Yoshihara, S. Ito, and Y. Isozaki for collecting the BIF samples; S. Yogo for preparing the thin sections; and T. Goto for technical advice on the SEM-EDS analysis. This study was supported by the Decoding Earth Evolution Program (DEEP) Grant-in-Aid for Scientific Research on Priority Areas (No. 07238104); a Grant-in-Aid for Scientific Research for

Young Scientists (B) (No. 19740319, 24700947); Dynamics of the Sun-Earth-Life Interactive System, Number G-4, the 21st Century COE Program for the Ministry of Education, Culture, Sports, Science and Technology, Japan; the Saijiro Endo Memorial Foundation; and the Fujiwara Natural History Foundation.

Author details

Nagayoshi Katsuta[1], Ichiko Shimizu[2], Masao Takano[3], Shin-ichi Kawakami[1], Herwart Helmstaedt[4] and Mineo Kumazawa[3]

1 Faculty of Education, Gifu University, Japan

2 Department of Earth and Planetary Science, Graduate School of Science, University of Tokyo, Japan

3 Graduate School of Environmental Studies, Nagoya University, Japan

4 Department of Geological Science, Queen's University, Canada

References

[1] James HL. Sedimentary facies of iron-formation. Economic Geology 1954;49(3) 235–293.

[2] Klein C, Beukes NJ. Time distribution, stratigraphy, and sedimentologic setting, and geochemistry of Precambrian iron-formations. In: Shopf JW, Klein C. (ed.) The Proterozoic Biosphere: A Multidisciplinary Study. New York: Cambridge University Press; 1992. p139–146.

[3] Isley AE, Abbott DH. Plume-related mafic volcanism and the deposition of banded iron formation. Journal of Geophysical Research 1999;104 (B7) 15461–15477.

[4] Trendall AF, Blockley JG. The iron formations of the Precambrian Hamersley Group, Western Australia. Geological Survey of Western Australia Bulletin 1970;119 1–366.

[5] Trendall AF. Varve cycles in the Weeli Wolli Formation of the Precambrian Hamersley Group, Western Australia. Economic Geology 1973;68(7) 1089–1097.

[6] Walker JCG, Zahnle KJ. Lunar nodal tide and distance to the Moon during the Precambrian. Nature 1986;320(6063) 600–602.

[7] Williams GE. Geological constraints on the Precambrian history of Earth's rotation and the Moon's orbit. Reviews of Geophysics 2000;38(1) 37–59.

[8]   Simonson BM, Hassler SW. Was the deposition of large Precambrian iron formations linked to major marine transgressions? Journal of Geology 1996;104(6) 665–675.

[9]   Pickard AL, Barley ME, Krapež B. Deep-marine depositional setting of banded iron formation: sedimentological evidence from interbedded clastic sedimentary rocks in the early Palaeoproterozoic Dales Gorge Member of Western Australia. Sedimentary Geology 2004;170(1–2) 37–62.

[10] Klein C. Some Precambrian banded iron-formations (BIFs) from around the world: Their age, geologic setting, mineralogy, metamorphism, geochemistry, and origin. American Mineralogist 2005;90(10) 1473–1499.

[11] Böning P, Bard E, Rose J. Toward direct, micron-scale XRF elemental maps and quantitative profiles of wet marine sediments. Geochemistry Geophysics Geosystem 2007;8(5): Q05004.

[12] Katsuta N, Takano M, Kawakami SI, Togami S, Fukusawa H, Kumazawa M, et al. Climate system transition from glacial to interglacial state around the beginning of the last termination: Evidence from a centennial- to millennial-scale climate rhythm. Geochemistry Geophysics Geosystem 2006;7(12): Q12006.

[13] Katsuta N, Takano M, Kawakami SI, Togami S, Fukusawa H, Kumazawa M, et al. Advanced micro-XRF method to separate sedimentary rhythms and event layers in sediments: its application to lacustrine sediment from Lake Suigetsu, Japan. Journal of Paleolimnology 2007;37(2) 259–271.

[14] Katsuta N, Tojo B, Takano M, Yoshioka H, Kawakami S, Ohno T, et al. Non-destructive method to detect the cycle of lamination in sedimentary rocks: rhythmite sequence in Neoproterozoic cap carbonate. In: Vickers-Rich P, Komarower P. (ed.) The Rise and Fall of the Ediacaran Biota. Geological Society, London, Special Publication. Bath: The Geological Society Publishing House; 2007. p27-34.

[15] Kuroda J, Ohkouchi N, Ishii T, Tokuyama H, Taira A. Lamina-scale analysis of sedimentary components in Cretaceous black shales by chemical compositional mapping: Implications for paleoenvironmental changes during the Oceanic Anoxic Events. Geochimica et Cosmochimica Acta 2005;69(6) 1479–1494.

[16] Sakai H, Shirai K, Takano M, Horii M, Funaki, M. Analysis of fine structure of chert and BIF by measurement of high resolution magnetic field and scanning X-ray analyzed microscope. Proceedings of the NIPR symposium on Antarctic Geosciences 1997;10 59–67.

[17] Matsunaga M, Fukuda K, Kato Y, Nakai I. Characterization of banded iron formations by two-dimensional XRF imaging and XANES analyses. Resource Geology 2000;50(1) 75–81.

[18] Fukuda K, Matsunaga M, Kato Y, Nakai I. Chemical speciation of trace titanium in Hamersley banded iron formations by X-ray fluorescence imaging and XANES analysis. Journal of Trace and Microprobe Techniques 2001;19(4) 509–519.

[19] Pufahl PK, Fralick PW. Depositional controls on Palaeoproterozoic iron formation accumulation, Gogebic Range, Lake Superior region, USA. Sedimentology 2004;51(4) 791–808.

[20] Katsuta N, Shimizu I, Helmstaedt HH, Takano M, Kawakami S, Kumazawa M. Major element distribution in Archean banded iron-formation (BIF): Influence of metamorphic differentiation. Journal of Metamorphic Geology 2012;30(5) 457–472.

[21] MacLachlan K, Helmstaedt H. Geology and geochemistry of an Archean mafic dike complex in the Chan Formation: basis for a revised plate–tectonic model of the Yellowknife greenstone belt. Canadian Journal of Earth Sciences 1995;32(5) 615–630.

[22] Isachsen CE, Bowring SA. The Bell Lake group and Anton Complex: a basement–cover sequence beneath the Archean Yellowknife greenstone belt revealed and implicated in greenstone belt formation. Canadian Journal of Earth Sciences 1997;34(2) 169–189.

[23] Mueller WU, Corcoran PL, Pickett C. Mesoarchean continental breakup: evolution and inferences from the >2.8 Ga Slave craton–cover succession, Canada. Journal of Geology 2005;113(1) 23–45.

[24] Ketchum J, Bleeker W. New field and U-Pb data from the Central Slave Cover Group near Yellowknife and the Central Slave Basement Complex at Point Lake. In: Cook F, Erdmer P. (ed.) Slave-Northern Cordillera Lithospheric Evolution (SNORCLE) Transect and Cordilleran Tectonic Workshop Meeting, February 25–27, 2000, University of Calgary, Canada. Lithoprobe Report 72;2000. p27–31.

[25] Joint Committee on Power Diffraction Standards. Mineral Powder Diffraction File Data Book: Sets 1–50. International Centre for Diffraction Data 2001.

[26] Droop GTR. A general equation for estimating Fe3+ concentrations in ferromagnesian silicates and oxides from microprobe analysis, using stoichiometric criteria. Mineralogical Magazine 1987;51(361) 431–435.

[27] Hosokawa Y, Ozawa S, Nakazawa H, Nakayama Y. An x-ray guide tube and a desktop scanning x-ray analytical microscope. X-Ray Spectrometry 1997;26(6) 380–387.

[28] Katsuta N, Takano M, Okaniwa T, Kumazawa M. Image processing to extract sequential profiles with high spatial resolution from the 2D map of deformed laminated patterns. Computers & Geosciences 2003;29(6) 725–740.

[29] Leake BE, Woolley AR, Arps CES, Birch WD, Gilbert MC, Grice JD, et al. Nomenclature of amphiboles; report of the Subcommittee on Amphiboles of the International Mineralogical Association Commission on New Minerals and Mineral Names. European Journal of Mineralogy 1997;9(3) 623–651.

[30] Togami S, Takano M, Kumazawa M, Michibayashi K. An algorithm for the transfor-
      mation of XRF images into mineral-distribution maps. Canadian Mineralogist
      2000;38(5) 1283–1294.

[31] Goryainov PM, Ivanyuk GYu, Sharov NV. Fractal analysis of seismic and geological
      data. Tectonophysics 1997;269(3–4) 247–257.

[32] Morris RC. Genetic modelling for banded iron-formation of the Hamersley Group,
      Pilbara Craton, Western Australia. Precambrian Research 1993;60(1–4) 243–286.

[33] Rothman DH, Grotzinger JP, Flemings P. Scaling in turbidite deposition. Journal of
      Sedimentary Research 1994;A64(1) 59–67.

[34] Longhitano SG, Nemec W. Statistical analysis of bed-thickness variation in a Tortoni-
      an succession of biocalcarenitic tidal dunes, Amantea Basin, Calabria, southern Italy.
      Sedimentary Geology 2005;179(3–4) 195–224.

[35] Kretz R. Symbols for rock-forming minerals. American Mineralogist 1983;68(1-2)
      277–279.

Chapter 8

Trace Elements and Palynomorphs in the Core Sediments of a Tropical Urban Pond

Sueli Yoshinaga Pereira, Melina Mara de Souza,
Fresia Ricardi-Branco, Paulo Ricardo Brum Pereira,
Fabio Cardinale Branco and
Renato Zázera Francioso

Additional information is available at the end of the chapter

1. Introduction

Park Hermogenes de Freitas Leitao Filho is located in Campinas (SP), limited by the Cidade Universitaria I and Cidade Universitaria II neighbourhoods and the State University of Campinas (UNICAMP), in the Barão Geraldo district, whose population was estimated in 2011 to be around 60,000 inhabitants [1]. The park has an estimated area of 123,901.06 m^2 and of this total, 80,111.17 m^2 corresponds to the surface of a pond formed by the damming of two streams: one passes through the campus of UNICAMP, draining an area of 325.813 m^2. In recent decades, effluent from the University has been released into the pond. In 2004, this release was captured by the sewage system of the municipal sanitation company. Currently, the pond receives urban drainage water from the Cidade Universitária II neighbourhood and a Centro Médico Hospital, on the right bank, and the university campus and the Cidade Universitaria II neighbourhood on the left bank.

Thus, in this study we sought the presence of trace elements (As, Co, Cr, Cu, Ga, Ni, Pb, Th, V and Zn) and Al_2O_3, Fe_2O_3, MnO and Loss on Ignition (LOI, 105 ºC and 1,000 ºC) in recent sediments of the pond and correlate it with the occurrence of pollen and spores derived from the surrounding vegetation. Accordingly, a 65 cm-deep core, named T-UNICAMP, forming the subject matter of this article.

2. Study area

Park Hermogenes de Freitas Leitão Filho is located at the coordinates 22°48'40.20" S and 47°04'11.86" W (23k, 287000 to 287700 - 7475700 to 7476000 m) and it is one of the green leisure areas surrounding the University of Campinas, Campinas (SP) in the district of Barão Geraldo.

The park and its surroundings have undergone significant changes in their vegetation cover and land use from the 1950s onwards, mainly because of the growth of the neighbourhood and the District of Barão Geraldo, as well as construction on the university campus and the urban expansion of the city of Campinas. The land use has changed from wood savannah (Brazilian Cerrado) to agricultural use (sugar cane crops and pasture), and eventually to urban use (residential and the university).The surrounding vegetation is represented by degraded fragments of wood savannah (Brazilian Cerrado) and garden vegetation.

Figure 1 shows the study area and the location of the analysed sample.

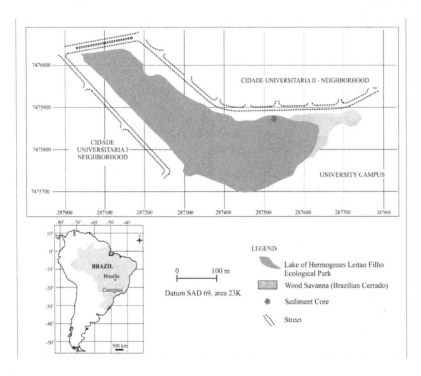

Figure 1. The study area and localization of the sample core

3. Material and method

The 65 cm-depth core (T-UNICAMP) was removed near the pluvial exit of the Cidade Universitaria II neighbourhood. The material collected was predominantly sandy-clay.

For the chemical analysis of the sediments, samples were taken every 10 cm, for a total of six samples. In these samples, the elements As, Co, Cr, Cu, Ga, Ni, Pb, Th, V and Zn, and Al_2O_3, Fe_2O_3, MnO were analysed by X-ray fluorescence spectrometry (Philips PW 2404) and the Loss on Ignition (105°C and 1000°C) parameters at the Analytical Geochemistry Laboratory of the Geosciences Institute - UNICAMP. The larger elements were determined in the Uniquant program and the other elements were determined in Solo2007 – a Superq program.

For pollen analysis, six samples were collected at the same intervals of depth for the analysis mentioned above. The palynology samples were processed according to the classical method of [2] for Quaternary sediments, which comprises the following steps: dissolution of silicates by HF; removal of silica colloidal with diluted HCl (hot); destruction of humic acids by 10% KOH solution; centrifugation and washing the sample with distilled water; blade mounting, with 50 microlitres of the sample, and observation under a Axioimager Zeiss microscope at the Palaeo Hydrogeology Laboratory of the Institute of Geosciences - UNICAMP.

4. Results

4.1. Description of sediments

The sediments are predominantly sandy clay, for the interval 0 to 40 cm deep. The sediments then become silty clay, up to 51.5 cm. From this depth, fine sandy sediments occur. Note that the sediments are oxidized to a depth of 30 cm.

Figure 2 presents a description of the sediments, the distribution of the palynomorphs and the chemical elements along the sampled profile.

4.2. Chemical analysis

The results of sediment analysis are presented in Table 1.

Depth	10 cm	20 cm	30 cm	40 cm	50 cm	65 cm
Al_2O_3 (%)	21.1	22.8	25.4	26.2	23.7	2.1
Fe_2O_3 (%)	2.4	2.5	1.3	1.2	15.3	0.3
MnO (%)	0.02	0.01	0.01	0.01	0.09	0.01
LOI (105°C) (%)	4.7	4.4	5.4	4.6	2.7	0.2
LOI (1000°C) (%)	20.2	17.2	17.0	14.0	12.9	0.9

Depth	10 cm	20 cm	30 cm	40 cm	50 cm	65 cm
As (ppm)	2.7	3.3	1.0	1.0	5.1	1.0
Co (ppm)	8.8	10.6	6.4	7.9	20.0	8.0
Cr (ppm)	77	78	202	204	77	18
Cu (ppm)	82	8	11	152	166	6
Ga (ppm)	26.0	27.0	33.0	32.0	30.0	4.5
Ni (ppm)	25.0	27.0	34.0	36.0	42.0	3.2
Pb (ppm)	29.0	24.0	34.0	41.0	26.0	6.1
Th (ppm)	11.6	12.3	16.0	15.7	10.5	2.9
V (ppm)	227	223	103	94	359	23
Zn (ppm)	118	380	57	40	171	6

LOI – Loss On Ignition

Table 1. Results of the chemical analysis of sediments.

The Al_2O_3 concentrations varied from 21.1 to 26.2% to a depth of 40 cm and a minimum value of 2.1 % on the basis of the core. The Fe_2O_3 concentration was found between 0.3 and 15.3 % at depths of 65 and 50 cm respectively. In the other depths, the concentration ranged from 1.2 to 2.5 %.

The MnO has concentrations ranging from 0.01 to 0.09%, which is highest in the sample at 50 cm deep.

The LOI values ranged from 0.2 to 5.4% (LOI 105°C) and 0.90 to 20.20 % (LOI 1000°C). The lowest values were found at the base of the core.

Regarding the trace elements, the concentrations of As ranged from 1.0 to 5.1 ppm, Co from 6.4 to 20.0 ppm, Cr from 18 to 204 ppm. The concentrations of Cu, Ga and Ni ranged from 6 to 166 ppm, 4.5 to 33.0 ppm and 3.2 to 42.0 ppm, respectively. The elements Pb, Th, V and Zn presented concentrations ranging from 6.1 to 41.0 ppm, 2.9 to 16.0 ppm, 23 to 359 ppm and 6 to 380 ppm, respectively.

On the basis of the core (interval from 65 to 50 cm), the lowest concentrations of the trace elements analysed are found in the most sandy portion.

In the 51.5 - 40 cm and 40 - 30 cm intervals, there is a higher proportion of clay in the reduced environment with organic matter, which favours the retention of trace elements such as Cr, Cu, Ga, Ni, Pb and V. The depth interval from 50 to 40 cm has higher concentrations of the chemical elements As, Co, Cu, Ni, V and Zn. The 20 - 0 cm interval presented a high concentration of Zn and significant concentrations of other elements.

The Al_2O_3 concentration due to the presence of clay has little variation (between 21.1 to 26.2 %) up to the 65 cm interval, which has the lowest value (2.1%) in the sandy portion. The Fe_2O_3 concentration is found in highest percentage in the interval of 50 cm, along with MnO.

The Loss on Ignition parameter (105°C and 1000ºC) occurs in all the intervals, ranging from 0.2 to 5.4% and 0.9 to 20.2%, respectively. Thus, organic matter is present in the more superficial portions of the core to a depth of 50 cm, with a strong decrease in the sandy portion (65 cm interval).

4.3. Pollen analysis

As it can be seen in Figure 2, pollen grains were identified in two intervals: T-UNICAMP/32-35 cm and T-UNICAMP/42-46 cm. Figure 2 shows, in alphabetical order, the main pollen types distribuition found, of which 19 were Angiosperms and 3 were Pteridophytes (Figure 3). We observed the existence of 446 pollen grains distributed in 19 pollen types and 36 spores. Among the different taxonomic categories we considered as indeterminate those that could not be identified to family level, which amounted to 41 grains of pollen.

The 42-46 cm interval is mainly characterized by a low concentration of pollen grains and the presence of spores. In this interval, we mainly observed the presence of Cyperaceae, Aralia-ceae, Poaceae and Myrtaceae. As for the spores, these were of the families Polypodiaceae *(Polypodium)* and Cyatheaceae *(Cyathea).*

The 32-35 cm interval is characterized by a higher concentration and amount of pollen grains and the presence of rate indicators of wood savannah (Brazilian Cerrado). However, we did not observe a significant number of Pteridophytes spores. The predominant families in this interval were: Araliaceae, Asteraceae, Rubiaceae, Malpighiaceae, Myrtaceae, Fabaceae and Cyperaceae - featuring wood savannah (Brazilian Cerrado) - present in the remaining frag-ments of the vegetation surrounding the pond. As for the spores, we saw the presence of Cyatheaceae, Polypodiaceae and Dicksoniaceae.

5. Discussion

Urban lakes suffer pollution problems arising primarily from the activities of their urban surroundings, which may contribute to the greater amount of trace elements, which are concentrated in the pond fine sediments. Trace elements such as Cd, Cu, Pb and Zn are toxic and present adversity to aquatic organisms and humans [3].

The elements As, Cr, Cu, Ni and Zn (T.E.) were analysed taking as sediment toxicity screening values for aquatic life the TEL and PEL - according to [4] - indices adopted by the Sao Paulo state environmental agency. The acronym TEL means "Threshold Effect Level" and PEL means "Probable Effect Level". These values are guidelines for sediment quality, and differ in values for each parameter analysed, although both aim to protect life in aquatic environments. The proposed values of these bands are divided into three parts: below the minimum value suggested, where an adverse effect is rarely expected (<TEL); between the minimum and the maximum value, where the possibility of an adverse effect might be expected (> TEL and <PEL); and the higher than the maximum value suggested (> PEL), where an adverse effect is often expected. Thus, Table 2 presents the patterns of TEL and PEL for each element (As, Cr, Cu, Ni and Zn), the depths of their occurrence and their classification as to the quality of the sediments.

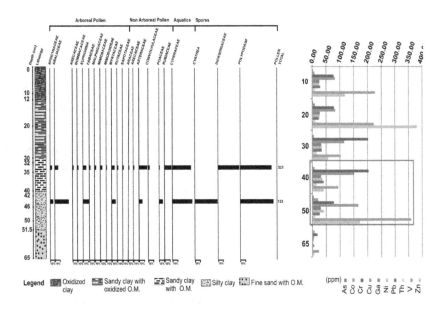

Figure 2. Descriptive profile of the T-UNICAMP core featuring a pollen diagram and the distribution of trace elements (T.E.) in depth.

T.E.	TEL (ppm)	PEL (ppm)	<TEL (depth)	"/>TEL and <PEL (depth)	"/>PEL (depth)
As	5.9	17	10, 20, 30, 40, 50 and 65 cm	-	-
Cr	37.3	90	65 cm	10, 20, 50 cm	30 and 40 cm
Cu	35.7	˙ 197	65 cm	10, 20, 30, 40 and 50 cm-	
Ni	18	36	65 cm	10, 20, 30 and 40 cm	50 cm
Zn	123	315	10, 30, 40 and 65 cm	50 cm	20 cm

Table 2. TEL and PEL results from a comparative analysis of the trace elements' (T.E.) distribution in depth.

The most adverse concentrations of Cr (> PEL) are found in depths of 30 and 40 cm. At the depth of 50 cm, Ni concentrations exceed the PEL index and at the depth of 20 cm, the element Zn is greater than the PEL index. Above the TEL index, but below the PEL, samples from depths of 10, 20, 30 and 40 cm present Cu and Ni elements within this range of values. Nonetheless, within this PEL / TEL interval, at the depth of 50 cm, there are Ni and Zn elements.

In the sand fraction at the greater sampled depth (65 cm), we found lower concentrations of the elements studied (<TEL).

The element As was found in concentrations below the TEL, where an adverse effect is rarely expected. Figure 2 shows the pollen diagram compared with the results of chemical analysis of the sediments. Thus, in the intervals 42 - 46 cm and 32 - 35 cm, the occurrence and preservation of palynomorphs is related to high levels of trace elements like As, Co, Cr, Cu, Ni, Pb, V and Zn. This association may be related to an environment of reduced deposition [5, 6] where high concentrations above environmental standards - especially of Cr, Ni and Zn - present toxicity to those microorganisms and invertebrates that feed on organic matter.

The presence of fine sediments (clay and silt) and organic matter which may come from biomass killed by this toxicity favours the concentration of trace elements found in the depths [3, 7].

In the intervals sampled above 30 cm, the trace elements had concentrations above the environmental standards. However, oxidized sediments are presented by varying the water level of the pond, which would explain the non-preservation (absence) of palynomorphs in the upper levels of the core.

At intervals below 50 cm, the absence of palynomorphs and the low concentration of trace elements is a consequence of the occurrence of sandy sediments, indicating an environment of higher energy; as such, it is more abrasive for palynomorphs and has a lower capacity to absorb the elements studied [5, 6].

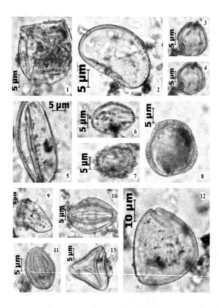

Figure 3. Palynomorphs: 1) Apocynaceae; 2) Araceae; 3) Araliaceae; 4) Araliaceae; 5) Arecaceae; 6) Bombacaceae; 7) Asteraceae; 8) Convolvulaceae; 9) Cyperaceae; 10) Euphorbiaceae; 11) Fabaceae; 12) Mimosoideae; and 13) Myrtaceae.

6. Conclusion

The urban lake of Park Hermogenes Leitão Filho has sediments with adverse registers for the elements Cr, Ni and Zn, possibly due to sewage discharge from urban occupation and services surrounding the pond. The presence of clay and organic matter contributed to a higher retention of these elements and palynomorphs in the sediments.

The study showed that in those areas strongly impacted by human activities the concentration of toxic elements in fine and anoxic sediments, the preservation of the assemblages of palynomorphs may occur, since this polluted environment does not allow for the survival of microorganisms and invertebrates that feed on organic matter.

Author details

Sueli Yoshinaga Pereira[1], Melina Mara de Souza[1], Fresia Ricardi-Branco[1],
Paulo Ricardo Brum Pereira[2], Fabio Cardinale Branco[3] and Renato Zázera Francioso[1]

*Address all correspondence to: fresia@ige.unicamp.br

1 Institute of Geosciences, State University of Campinas, Brazil

2 Forestry Institute, Sao Paulo State Environmental Secretariat, Brazil

3 EnvironMentality – Conceitos Ambientais LTDA, Brazil

References

[1] Instituto Brasileiro de Geografia e Estatística (IBGE) Censo Demográfico 2000. http://www.ibge.gov.br/english/estatistica/populacao/default_censo_2000.shtm[Retrieved-September 20, (2011).

[2] Faegri, K, & Iversen, J. (1989). *Textbook of Pollen Analysis.* The Blackburn Press. 4th Ed. 328p.

[3] Das, S. K, Routh, J, & Roychoudhury, A. N. (2008). Major and trace element geochemistry in Zeekoevlei, South Africa: A lacustrine record of present and past processes. *Applied Geochemistry.* 23. , 2496-2511.

[4] Canadian Council of Ministers of the Environment (CCME) *Environmental Quality Guidelines- sediment quality guidelines for the protection of aquatic life. Summary Table.* http://st-ts.ccme.ca/,Retrieved October 13, (2011).

[5] Lebreton, V, Messager, E, Marquer, L, & Renault-Miskovsky, J. (2010). A neotaphonomic experiment in pollen oxidation and its implications for archaeopalynology. *Review of Palaeobotany and Palynology*. 162. , 29-38.

[6] Twiddle, C. L, & Bunting, M. J. (2010). Experimental investigations into the preservation of pollen grains: A pilot study of four pollen types. *Review of Palaeobotany and Palynology*. 162. , 621-630.

[7] Bilali, L. E, Rasmussen, P. E, Hall, G. E. M, & Fortin, D. (2002). Role of sediment composition in trace metal distribution in lake sediments. *Applied Geochemistry*. 17, , 1171-1181.

Chapter 9

Mapping of Lineaments for Groundwater Targeting and Sustainable Water Resource Management in Hard Rock Hydrogeological Environment Using RS- GIS

Pothiraj Prabu and Baskaran Rajagopalan

Additional information is available at the end of the chapter

1. Introduction

Lineament definition and history

Numerous definitions of the term 'lineament' are given in the literature and various attributes are sometimes linked to the term - such as 'geologic lineament', 'tectonic lineament', 'photo lineament' or 'geophysical lineament' - either describing the assumed origin of the linear feature or sometimes the data source from which it has been derived. Some researchers also use the term 'fracture trace' or 'photo linear' as an alternative term. The work by Lattman and Parizek (1964) is commonly regarded as pioneering work in groundwater exploration; they mapped linear features (fracture traces) on stereo-pairs of aerial photographs in carbonate terrain in the eastern United States and subsequently showed the correlation between well productivity and distance to the identified features.

Lineament mapping was used long before this work in other geological applications and the first usage of the term lineament in geology is probably from a paper by Hobbs (1904, 1912), who defined lineaments as significant lines of landscape caused by joints and faults, revealing the architecture of the rock basement. This was later used by O' Leary et al. (1976) as a basis for developed definitions. Lineaments have been defined as extended mappable linear or curvilinear features of a surface whose parts align in straight or nearly straight relationships that may be the expression of folds, fractures or faults in the subsurface. These features are mappable at various scales, from local to continental, and can be utilized in mineral, oil and gas, and groundwater exploration studies.

Linear features on the Earth's surface have attracted the attention of geologists for many years. This interest has grown most rapidly in geological studies since the introduction of aerial photographs and satellite images. At the beginning, to the middle of the twentieth century, several geologists recognized the existence and significance of linear geomorphic features that were the surface expression of zones of weakness or structural displacement in the crust of the Earth.

Studies revealed a close relationship between lineaments and groundwater flow and yield (Mabee et al., 1994; Magowe and Carr, 1999; Fernandes and Rudolph, 2001). Generally lineaments are underlain by zones of localized weathering and increased permeability and porosity. Meanwhile, some researchers studied relationships between groundwater productivity and the number of lineaments within specifically designated areas or lineament density rather than the lineament itself (Hardcastle, 1995). Therefore, mapping of lineaments closely related to groundwater occurrence and yield is essential to groundwater surveys, development and management. In the last two decades remote sensing and GIS have been widely used for preparation of different types of thematic layers and their integration for different purposes.

This research work focuses on developing the remote sensing and Geographic Information Systems (GIS) methodology for regional groundwater potential evaluation. The objectives of this study are to (i) produce a regional structural lineament map of the study area from remotely sensed data, (ii) determine the hydro geological implication of the lineaments by integrating them with the available ancillary data (Digital Elevation Model [DEM] and geological map), (iii) analyse the lineament trend distribution of the study area using rose diagrams, lineament density maps and lineament intersection maps.

2. Description of study area

The Vaigai sub-basin extends over approximately 849 km^2 and lies between 09^0 30` 00'' and $10^0$ 00` 00''N latitudes and 77^0 15` 10''and $77^0$ 45 00` E longitudes in the western part of Tamilnadu, India. It originates at the altitude of 1661m in the Western Ghats of Tamilnadu in the Theni district (Figure 1). The basin is generally hot and dry except during the winter season. The maximum and minimum temperature for the basin is 40.7 ^{0}C and 16.0 ^{0}C. The area receives an average annual rainfall of about 384 mm. The surface runoff goes to stream as instant flow. Rainfall is the direct recharge source and the irrigation return flow is the indirect source of groundwater in the Vaigai sub-basin. The study area depends mainly on the north-east monsoon rains which are brought by the troughs of low pressure established in the Bay of Bengal. Most of the farmers depend on the groundwater for their irrigational needs. There are a few tanks across these drainages, however, most of these remain dry.

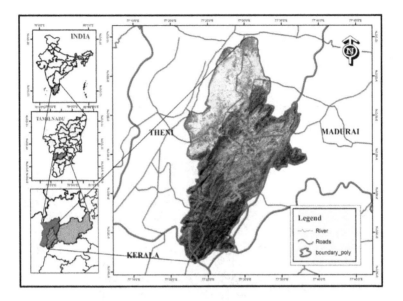

Figure 1. Location of the study area with ASTER 30m DEM

3. Methodology

The Indian Remote Sensing Satellite (IRS) ID, Linear Image Self-Scanning (LISS) III of geo-coded False Colour Composites (FCC), generated from the bands 2, 3 and 4 on 1:50,000 scale was used for the present study. The application of higher-resolution 30- m Advanced Space borne Thermal Emission and Reflection Radiometer (ASTER) imagery yielded better results in lineament interpretation compared to IRS 1D imagery due to improved spatial resolution. Lineament mapping is normally undertaken based on geomorphological features such as aligned ridges and valleys, displacement of ridge lines, scarp faces and river passages, straight drainage channel segments, pronounced breaks in crystalline rock masses and aligned surface depression For the study area, the main interest was topographically negative lineaments, which may represent joints, faults and probably shear zones (Juhari and Ibrahim 1997; Koch and Mather 1997; Solomon and Ghebreab 2006). To eliminate the non-geological elements, such as paths, roads, power cables and field boundaries in the study area, geographical maps and field checking were undertaken using the method suggested by Yassaghi (2006).

3.1. Geology

Eleven geologic features were identified and mapped by the Geological Survey of India, shown in Figure 2.

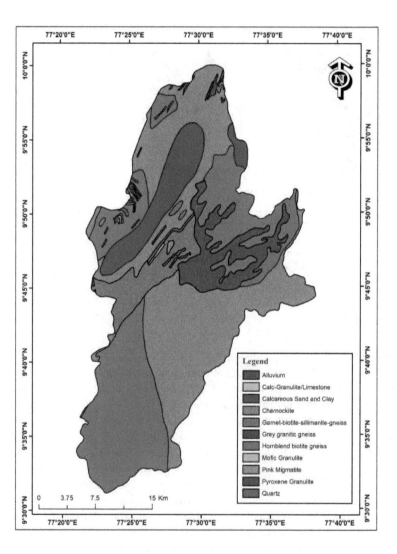

Figure 2. Geology of the study area

3.2. Lineament analysis

The mapped structural lineaments were mapped and analysed using the lineament density (LD), lineament frequency (LF) and lineament intersection (LI) parameters. The results of the analysis are presented as the lineament map, lineament density map, rose diagram, lineament frequency map and lineament intersection map (Figures 3, 4, 5, 6 & 7) respectively.

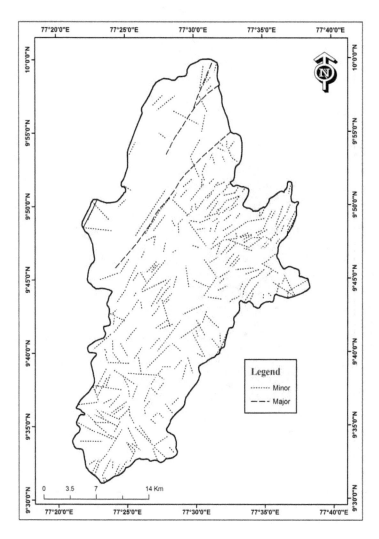

Figure 3. Lineament map of the study area

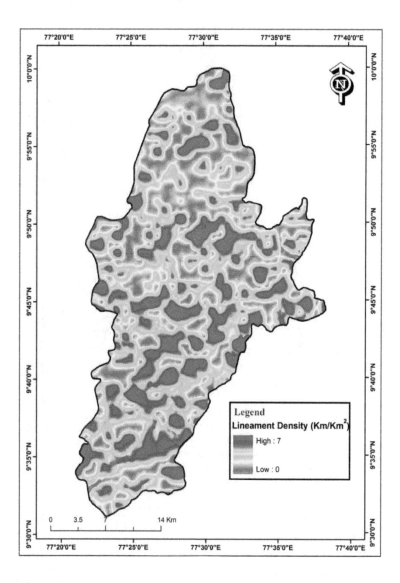

Figure 4. Lineament density of the study area

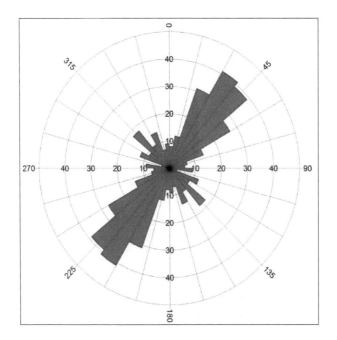

Figure 5. Lineament rose plot in the study area

4. Results and discussion

4.1. Frequency and spatial location of the fractures

The lineament and frequency map (Fig. 3 and Fig. 6) shows that the lineaments/fracture distribution is hardly homogeneous. The lineament density variation map (Fig. 6) shows the lineament numbers to be in the range of 0 and 7. The majority of the fractures are located on lithologies that correspond to the term "hard rocks", which generally refers to igneous and metamorphic rocks (Krasny 1996, 2002). Therefore, the discussed character represents an initial indication for the unified tectonic and hydrogeologic behaviour of the hard rock environment. The majority of the lineaments/fractures are located on the Hornblend biotite gneiss and minority of the lineaments/fractures are located on the Pink Migmatite (Table 1).

4.2. Orientation of the lineaments

The orientation of the lineaments is analysed by constructing rose diagrams (Fig.5). Even though these diagrams are not length-weighted, they can indicate on each occasion what the most dominant directions of the fractures are. This analysis is very critical for the study of groundwater flow, as in most cases the orientation of the fractures is identical to the orientation of the preferential flow path.

The faults rose plot indicates two sets of orientation classes. The main two classes have NE and SW strike, while others have NW and SE strike. The uniformity of fracture orientation becomes an additional indication for the hydrogeologic regime.

No	Description	Frequency	Percent (%)	Length (Km)	Percent (%)
1	Hornblend biotite gneiss	163	44.05	221.68	45.51
2	Charnockite	104	28.11	158.27	32.49
3	Alluvium	54	14.59	57.81	11.87
4	Garnet-biotite-sillimanite-gneiss	26	7.03	28.60	5.87
5	Calcareous Sand and Clay	7	1.89	14.17	2.91
6	Quartz	11	2.97	3.17	0.65
7	Pyroxene granulite	1	0.27	1.43	0.29
8	Grey granitic gneiss	2	0.54	1.16	0.24
9	Calc-granulite/Limestone	1	0.27	0.55	0.11
10	Pink migmatite	1	0.27	0.28	0.06
	Total	370	100.00	487.13	100.00

Table 1. Total length of lineaments in each geological feature

4.3. Size of the lineaments

Fracture dimensions (aperture and apparent aperture) are very difficult to define and the depth of the apertures makes the measurements even more complicated. Nevertheless, length measurements can be taken relatively easily and they are also significant, since a fracture with a greater length affects the groundwater flow in a more dominant way than those of smaller length. The calculated total length of lineament/fracture per unit area in each lithology are shown in Table 1.

4.4. Density of the lineaments

The purpose of the fracture density analysis is to calculate frequency of the fractures per unit area. With this analysis a map has been produced showing concentrations of the lineaments over the study area (Fig. 4). The map in Figure 3 shows that very high density is observed in areas of Hornblende biotite gneiss and Charnockite (7 Km/ Sq.km^2), indicates the high degree of hydraulic interconnection between the above lithologic units as surface water circulates through these discontinuities. This is verified in the next consideration (degree of fractures intersection). On the other hand, very low density is observed in calcareous sand and clay (1

Km/ Sq.km²), and quartz in areas where combination of more lithological features dominate. This verifies that these lithologies are affected by tectonic activity.

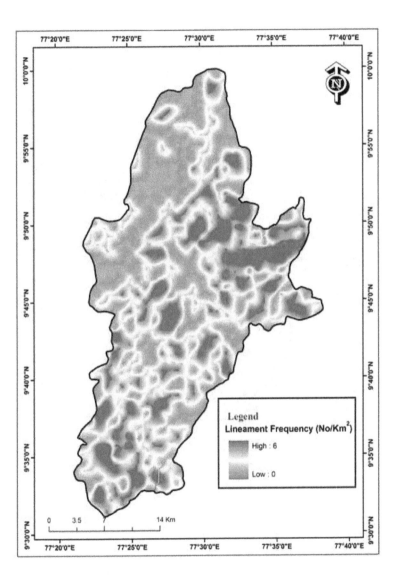

Figure 6. Lineament frequency of the study area

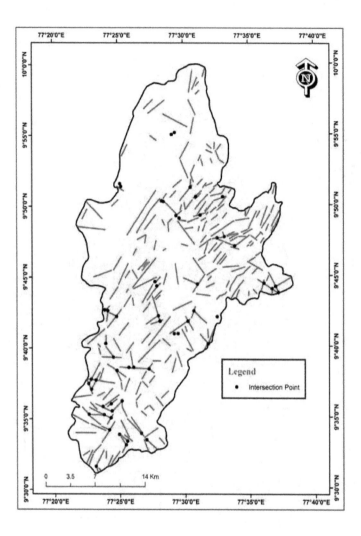

Figure 7. Lineament intersection of the study area

4.5. Degree of lineaments intersection

The density of lineaments along with the degree of lineament intersection determine the degree of anisotropy of groundwater flow in the fracture network, as in environments with a high degree of interconnection where groundwater flow is smoother and more uniform. Fracture intersection density is a map showing the frequency of intersections that occur in a unit area. The purpose of using intersection density maps is to estimate the areas of diverse fracture orientations. If the fractures do not intersect in an area, the resultant map will be represented by a plain map with almost no density contours and the fractures are almost parallel or sub-parallel in an area. The lineament intersection map of the study area (Fig. 7) indicates high and very high intersection in the same areas where there is very high density of lineaments. The zones of high lineament intersection over the study area are feasible zones for groundwater potential evaluation.

5. Conclusions

Groundwater studies on hard formations often require extraction of data from images and remote sensing, and GIS. Due to insufficient data, maps of lineament and structural elements are important tools that may reveal points of groundwater recharge and discharge, flow and development. In particular, groundwater occurrences in hard formations are mainly controlled by the lineaments corresponding to fractures, joints and faults. Furthermore, the distribution of lineament is closely related to groundwater discharge points and their concentration.

Remote sensing has proved to be a useful tool in lineament identification and mapping. This study demonstrates the application of remotely sensed data for lineament interpretation in a hard rock hydrogeological environment. A Digital Elevation Model (DEM) was generated to improve the interpretation. The lineament analysis has been effectively done in a GIS environment. Thematic maps, such as lineament frequency, lineament density and lineament intersection, were prepared using the interpolation technique.

The results from the study show that the remote sensing technique is capable of extracting lineament trends in an inaccessible tropical forest. The study has led to the delineation of areas where groundwater occurrences are most promising for sustainable supply, suggesting where further geophysical surveys can be concentrated. It is therefore suggested that the high lineament intersection and density should be combed with detailed geoelectrical surveys for quantitative evaluation of the groundwater potential of the study area. Properly sited wells in drought-stricken areas could change the lives of many and the remote-sensing analysts and lineament interpreters around the world are without doubt important in this process.

Author details

Pothiraj Prabu and Baskaran Rajagopalan

Department of Industries and Earth Sciences, Tamil University, Thanjavur, Tamilnadu, India

References

[1] Fernandes, A. J, & Rudolph, D. L. (2001). The influence of Cenozoic tectonics on the Groundwater production capacity of fractured zones: A case study in Sao Paulo, Brazil. Hydrogeology Journal , 9, 151-167.

[2] Hardcastle, K. C. (1995). Photolineament factor: A new computer-aided method for remotely sensed fractured. Photogrammetric Engineering & Remote Sensing 61 (6), 739-747.

[3] Hobbs, W. H. (1904). Lineaments of the Atlantic border region. Geological Society of America Bulletin , 15, 483-506.

[4] Hobbs, W. H. (1912). Earth Features and Their Meaning: An Introduction to Geology for the Student and General Reader. Macmillan Co., New York, 347.

[5] Juhari, M. A, & Ibrahim, A. (1997). Geological Applications of Landsat Thematic Mapper Imagery: Mapping and Analysis of Lineaments in NW Peninsula Malaysia. ACRS. Available online at: www.gisdevelopment.

[6] Koch, M, & Mathar, P. M. (1997). Lineament mapping for groundwater resource assessment: a comparison of digital Synthetic Aperture Radar (SAR) imagery and stereoscopic Large Format Camera (LFC) photographs in the Red Sea Hills, Sudan. International Journal of Remote Sensing, , 27, 4471-4493.

[7] Krasny, J. (2002). Hard Rock Hydrogeology. 1st Workshop on Fissured Rocks Hydrogeology Proceedings, Athens, , 11-18.

[8] Krasny, J. (1996). Hydrogeological Environment in Hard Rocks: An attempt at its schematizing and terminological consideration. Acta Univesitatis Carolinae Geologica, , 40, 115-122.

[9] Lattman, L. H, & Parizek, R. R. (1964). Relationship between fracture traces and the occurrence of groundwater in carbonate rocks. Journal Hydrology , 2, 73-91.

[10] Mabee, S. B, Hardcastle, K. C, & Wise, D. U. (1994). A method of collecting and analyzing lineaments for regional-scale fractured-bedrock aquifer studies. Ground Water 32 (6), 884-894.

[11] Magowe, M, & Carr, J. R. (1999). Relationship between lineaments and ground water occurrence in western Botswana. Ground Water 37 (2), 282-286.

[12] Leary, O, Freidman, D. W, Pohn, J. D, & Lineaments, H. A. linear, lineation-some proposed new standards for old terms. Geological Society of America Bulletin , 87, 1463-1469.

[13] Solomon, S, & Ghebreab, W. (2006). Lineament characterization and their tectonic significance using Landsat TM data and field studies in the central highlands of Eritrea. Journal of African Earth Sciences, , 46, 371-378.

[14] Yassaghi, A. (2006). Integration of Landsat imagery interpretation and geomagnetic data on verification of deep-seated transverse fault lineaments in SE Zagrosa, Iran International Journal of Remote Sensing, , 27, 4529-4544.

Chapter 10

Itajuba yansanae Gen and SP NOV of Gnetales, Araripe Basin (Albian-Aptian) in Northeast Brazil

Fresia Ricardi-Branco, Margarita Torres,
Sandra S. Tavares, Ismar de Souza Carvalho,
Paulo G. E. Tavares and Antonio C. Arruda Campos

Additional information is available at the end of the chapter

1. Introduction

This paper provides a description of the morphology and anatomy of a fertile fossil, related to gnetalean lineage, which has been named *Itajuba yansanae*. A conclusion has been drawn regarding the paleoclimate when this taxon proliferated. It was collected in the Araripe basin in Brazil (Fig. 1), in the sedimentary rocks of the Santana Formation [1].

1.1. General considerations of paleoflora of the Crato Member of the Santana Formation

It is well known that during the deposition of the Crato Member, semi-arid paleoclimatic conditions prevailed in the northeast of Brazil and influenced the Araripe Basin [1, 2, 3, 4]. The presence of a system of lakes associated with the deposition of the Santana Formation may have favoured the maintenance of a more humid microclimate than the semi-arid conditions prevailing in the surrounding region, or at least a wetter season [5].

The paleoflora of the Santana Formation is famous around the world, since it represents one of the best-preserved records of the Aptian in tropical Gondwana [2, 3, 4, 5]. The assembly of macrofossils of this paleoflora [6] is composed of approximately 35% Pteridophytes, of the orders Filicales, Lycophytes, and Sphenophytes; 50% "gymnosperms" of the orders Gnetales, Coniferales, Cycadales, Bennettitales and some Pteridospermales [4, 7-15; among others] and 17% angiosperms related to the '*ANITA*' lines Magnoliids, Monocots and Eudicots [16-19]. The paleoflora may reflect the presence of forest near the margins of a lacustrine system, as well as aquatic macrophytes inhabiting the lake [2; 3, 17- 19]; moreover, the fossils may represent a succession of vegetation which grew there during the deposition of the bodies of limestone of the Crato Member.

2. Geologic location

The Araripe Basin, located between the states of Ceará, Piaui and Pernambuco in the Northeast of Brazil (Figure 1), is approximately 9,000 km² in area and 1,700m in width. Its geologic evolution is related to the fragmentation of the paleocontinent of Gondwana and the consequent opening of the southern Atlantic [1, 2].

In this study, we utilize the lithostratigraphic division proposed by Assine [1] for the Araripe basin, since it is the product of years of study by various authors, including the Brazilian Petroleum Industry (PETROBRAS), rather than the lithostratigraphic scheme proposed for the same layers by Martill [20]. Moreover, the stratigraphy is easily correlated with the geology of other Cretaceous basins in the northeast of Brazil (Parnaíba, Potiguar, Jatobá, Tucano, etc.).

The sedimentary rocks of the Aptian-Albian sequence of the Araripe Basin were deposited during a post-rift event which reactivated the subsidence of the area of the basin. This sequence is composed of the Barbalha Formation (lower portion) and Santana Formation (upper portion), which are most clearly exposed in the cliffs of the tableland of Araripe [1]. The Santana Formation represents the end of the second sedimentation cycle of the sequence, with an upward decrease in grain size, terminating with the deposition of the layered micritic limestone of the lower Crato Member. This limestone is found in discontinuous banks up to 60 meters in thickness, laterally interlinked with shales. At times, layers of gypsum are found above the limestone; these are known as the Ipubi beds. In the other locations, the sedimentary rocks of the Crato Member are in discordant contact with the upper member of the Santana Formation or Romualdo Member [1].

3. Materials and methods

The fossils studied consist of compressions permineralized with iron oxides, giving them a reddish brown colouration, which clearly distinguishes them from the micritic limestone matrix of the Crato Member. As in the case of the plant fossils described by Kunzmann et al. [4], only some branches of the fossils have preserved anatomical details. The specimens are associated with cranial and post-cranial fragments of small Osteichthes. The specimens studied (MPMA 30-0042.03 A and B) were collected from laminated layers of limestone in the Pedra Branca Quarry near Nova Olinda in the Brazilian state of Ceará, and constitute part of the scientific collection of the Paleontological Museum of Monte Alto "Prof. Antonio C. Arruda-Campos", in the municipality of Monte Alto, in the state of São Paulo in Brazil.

The morphological study of the specimens was made using an Axiocam 5.0 attached to a Zeiss Stemi SV6C stereomicroscope, and digital images of the fossil specimens were also registered

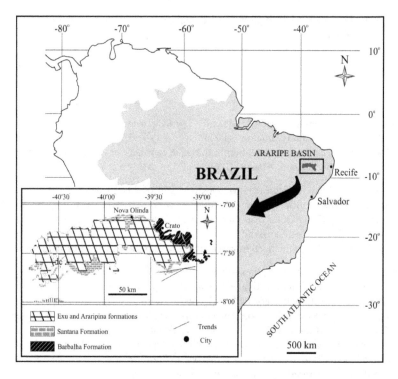

Figure 1. Location Map of the Araripe basin in Brazil.

with a Sony Alfa 1 camera (70mm lens). The lower part of specimen MPMA 30-0042.03 B was coated with gold and scanned using a LEO 430i Scanning Electron Microscope (SEM) of the Microscopic Laboratory of the Institute of Geosciences of UNICAMP and the JEOL-6360 Scanning Electron Microscope of the Institute of Chemistry of UNICAMP to obtain the longitudinal and cross section photographs.

The vessel density per mm^2 was measured for a random cross section, and the larger diameters of 50 vessels and tracheids were also measured.

In the present paper, the classification systems of the plant kingdom of Frey [21] and Kubitzki [22] were used.

4. Systematic palaeontology

Subkingdon Embryobionta Cronquist, Takhtajan and Zimmermann, 1966

Division Tracheophyta Sinnott, 1935 ex Cavalier-Smith, 1998

Figure 2. Photographs of external morphology of *I. yansanae*, showing vegetative characteristics: 1, overall view of sympodial branching, with terminal female cones and striated stems; 2, detail of View 1 showing striated stem (SS) and branching nodes (WB). Scale bars equal: 200mm in 1 and 100mm in 2.

Subdivision Spermatophytina Cavalier-Smith, 1998

Order Gnetales Luerssen, 1879

Genus *Itajuba* new genus

Figures 2, 3, 4, 5 and 6

Type species. *Itajuba yansanae*

Diagnosis. Plant with branch system bearing terminal female cones with striate stem at internodes. Main stem woody, with swollen nodes. Xylem consisting of vessels and tracheids, with tracheids more abundant than vessels; both with alternately distributed bordered pits. Long thin fibre-tracheids. Uniseriate rows of vascular rays composed of procumbent cells.

Reproductive shoot with ovulate cones in terminal branchlets, 1 ovulate/seed per cone, surrounded by pairs of bracts. Ovate seed with ornamented external surface.

Etymology. From Ita (stone), Juba (yellow) in Tupí-Guarani, since the laminated limestones of the Crato Member of the Santana Formation of the Araripe Group are yellow.

Itajuba yansanae new species

Figures 2, 3, 4, 5 and 6

Diagnosis. Sympodial branch system bearing terminal female cones with striate stem between all internodes with swollen nodes. Branches opposite-decussate. Xylem composed predominately of tracheids, with a few vessels (approximately 40 per mm²), both with helical thickenings; alternately distributed bordered pits. Long, thin fibre-tracheids. Uniseriate rows of vascular rays composed of procumbent cells. Reproductive shoot with female cones in terminal branchlets, 1 ovule/seed per cone, surrounded by two pairs of connate bracts. Ovate seed with ornamented external surface.

Description. Vegetative characteristics. Branches opposite-decussate, longitudinally striated and apparently leafless at maturity; sympodial branches (up to 6 orders) bearing organically connected female cones; more than 540mm long (Figures 2 and 3). Main axis woody, at least 225mm long and from 7.5 to 11.5mm wide with thickening at nodes.

Lower portion of branches thicker, at times preserving anatomical features of secondary xylem, such as vessels and tracheids. Following opposite-decussate branches considered to be of inferior orders. Second-order branches reach lengths of 145-205mm between nodes, with widths between 2.5 and 4mm. Third-order branches reach lengths of 56-125mm between nodes, with a width of 2mm. Fourth-order branches have a length of 6-56mm and a width of 1mm. Fifth-order branches 2-21mm long and 0.5mm wide, and those of the sixth orders bear organically connected female cones (Figures 3 and 4) and possibly ephemerous leaves.

Female reproductive structures. Fertile branches, jointed and longitudinally striated between whorls; 5.9-15.7mm in length. Female cones on terminal branches, 3.6-5.3mm in length and 2.6-2.8mm in width, enclosed by two pairs of sharply pointed bracts connated at the base and extending to or beyond the reproductive structure bearing a single ovule (Figures 3.1 – 3.4). Micropylar tube short and straight, 0.5 mm long. Ovate seed 3mm in length and 2.8mm in width; surface ornamented with apparent projections (Figures 3.6 and 4.1).

Anatomical characteristics. Cross-section. Composed of vessels and tracheids, with the latter much more abundant than the former, vessel density of 40 per mm² (Figure 5). At the widest point vessels of 22-55μm and tracheids 7-19μm. Walls of the vessels and tracheids with width of 1.5-4μm. Vascular. Uniseriate rows of vascular rays of procumbent cells. (Figure 5.1). Tangential section. Vessels and tracheids with helical thickenings (Figure 6). Borderer pits with wide, rounded openings (Figure 6.3) distributed alternately; thin fibre-tracheids, although not very clear, located around vessels and tracheids (Figure 6.7).

Etymology. From Yansan, female goddess of war governing spirits.

Holotype. MPMA 30-0042.03 A

Other materials examined. MPMA 30-0042.03 B

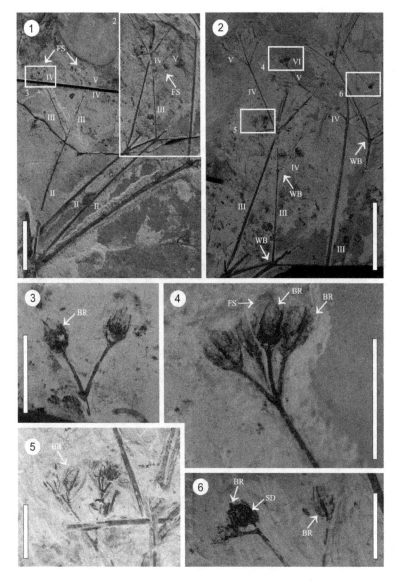

Figure 3. Photographs of external morphology of *I. yansanae* and female cones; 1, right-hand side branch bearing female cones (FS) organically connected to branch; 2, detail of View 1 showing reproductive structures and higher order branches (WB); 3, detail of stem of the sixth order bearing female cone organically connected and surrounded by

bracts (BR); 4, female cone (FS) showing pairs of bracts (BR) with connate bases and pointed tips; 5, female cone showing bracts (BR) protecting female reproductive structures; 6, detail of seed (SD) showing surface crosswise ornamentation with protuberances still surrounded by bracts (BR). Scale bars equal: 50mm in 1 and 10mm in 2,3,4,5 and 6.

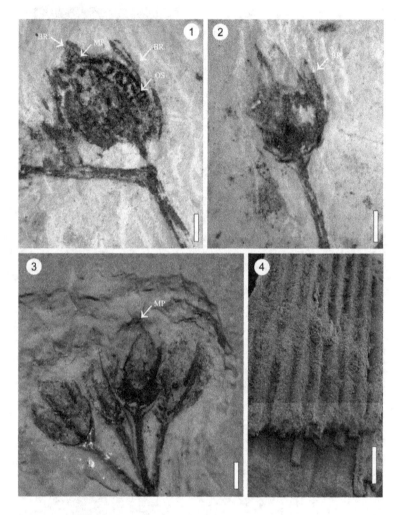

Figure 4. Photographs of external morphology of *I. yansanae* showing female cones and seed: 1, seed associated with
I. yansanae showing ornamented surface (OS), possible micropyle (MP) and bracts (BR); 2, female cone surrounded by
connate bracts (BR); 3, female cone showing bracts and a micropyle (MP); 4, SEM images of longitudinally striated
stem. Scale bars equal: 1mm in 1, 2, 3 and 4.

Occurrence. Between the urban centres of Nova Olinda and Santana do Cariri, in the Pedra Branca Quarry in the state of Ceará in Brazil. Lower level of the Crato Member (Aptian), of the Santana Formation, in the Araripe Basin.

Comparison and Discussion. To facilitate comparisons, two Tables (Tables 1 and 2) were elaborated, one for morphological aspects and the other for anatomy.

The specimens studied had an external morphology similar to the fossil taxa related to the Gnetales [23] including opposite-decussate branches longitudinally striated between nodes, terminal female cones, ovules/seed surrounded by bracts, seeds externally ornamented; as well as similar anatomical characteristics, such as the presence of vessels, tracheids and fibre-tracheids and the diameter of the vessels and tracheids.

In this order, the greatest similarity seems to be with *Ephedra*-like fossils, due to the striated stem between nodes, type of the female cone, and the presence of septate vessels, tracheids and fibro-tracheids, as well as the vascular rays with procumbent cells and the diameter of the vessels today found only in this group [24]. Based on the morphological and anatomical similarities, a new genus was proposed (*Itajuba* n. gen.) with the species designated *yansanae* n. sp. This new species was compared (see tables 1 and 2) with other gnetalean fossils found in lithostratigraphic units of the Lower Cretaceous in both the Northern and Southern hemispheres. It was compared with *Drewria potomacensis* Crane et Upchurch [25] from the Aptian of the Potomac Group of Virginia in the USA, and with *Ephedra archaeorhytidosperma* Yang *et al.* [26], *Liaoxia changii* (Cao et S.Q. Wu) Rydin, S.Q. et Friis [27], *L. chenii* Cao et S.Q. Wu [27], *Ephedra hongtaoi* Wang et Zheng [28] and *Siphonospermum simplex* Rydin et Friis [29], all species found in Barremian the Yixian Formation of Lianoning province in northeastern China. Moreover, *I. yansanae* was compared with fossil species related to the order Gnetales and *Ephedra* in the Lower Cretaceous found in the Southern Hemisphere: *Cearania heterophylla* Kunzmann *et al.* [4], *Cariria orbiculiconiformis* Kunzmann *et al.* [12], and a specimen possibly related to *Ephedra* [14], all collected in the same basin in Ceará in layers of the Crato Member of the Santana Formation (Aptian) in the Northeast of Brazil, as well as *Ephedra verticillata* Cladera *et al.* [30], found in the Ticó Formation of the Baqueró Group (Aptian) in Santa Cruz province in the south of Argentina.

Drewria potomacensis [25] had preserved leaves, and differences were found in the reproductive structures, which, although terminal for both, are loose spikes borne in dichasial groups of three in *D. potomacensis*, rather than consisting of terminal solitary female cones with one ovule/ seed.

Ephedra archaeorhytidosperma [26] shares the striated stem and single terminal female cones composed of 2 pairs of bracts, although the shape and size are different. Moreover, *E. archae-orhytidosperma* seems to have been a herb, whereas *I. yansanae* seems to have been a woody plant. Such differences would make it difficult to include the samples studied here in this genus.

A comparison of *I. yansanae* with *Liaoxia changii* and *L. chenii* shows that all three have striated stems and terminal female cones, but the bracts of the two species of *Liaoxia* are much larger;

moreover, the female cones in the Chinese species have from 2-10 pairs of bracts, whereas *I. yansanae* have only two.

Ephedra hongtaoi [28] was described to denominate a dioecious plant for which the roots, stems, branches and ovuliferous units are similar in gross morphology to *I. yansanae* with regard to the striated stem between nodes, reduced leaves and terminal female cones, although the morphology of the female cones is somewhat different and the anatomy is at present unknown.

Few comparisons can be established with *Siphonospermum simplex* [29]; although *S. simplex* and *I. yansanae* both have terminal reproductive units surrounded by bracts, the shape and size of these are different. Moreover, the former has a more developed micropylar tube.

Cearania heterophylla [4] has leaves, and the morphology of the reproductive units is also different (Table 1). The anatomical characteristics of the two are similar with regard to the presence of vessels, tracheids, and fibre-tracheids, as well as helical thickenings and alternate pits arranged in rows and longitudinally striated stems. The other species described by [12], designated *Cariria orbiculiconiformis*, may be related to the Gnetales, but it is also quite different from the species described here in terms of the morphology of the reproductive units and the presence of leaves (Table 1). The anatomical characteristics of *C. heterophylla* are similar in relation to the presence of the vessels and tracheids and pits, helical thickenings, although the cross-section of the xylem of both *C. heterophylla* and *Cariria orbiculiconiformis* is unknown, as well as the distribution and number of vessels per mm^2.

The specimen described by Fanton *et al.* [14] as possibly related to Gnetales is different from the species described here, especially with regard to size (much smaller) and the presence of opposite leaves and cones with more than two pairs of bracts. Both species do have a longitudinally striated stem.

E. verticillata was described for an impression/compression stems [30] has sessile seed-bearing structures either singly or in clusters of the three to five, whereas those of *I. yansanae* are uniformly singular.

The outer seed surface is profusely ornamented by rounded protuberations. Although due to the type of preservation of the fossils studied, this could not be observed in detail, this ornamentation resembles that mentioned by various authors [15, 26, 31-33]. Although not connected organically to the main stem, the seed was attached to a female cone identical to others, organically connected to the main branch. On the other hand, the seed associated with *I. yansanae* is clearly protected by bracts, as can be seen in Figures 3.2, 3.6 and 4.1.

5. Discussion and final remarks

The combination of morphological and anatomical characteristics makes a more complete interpretation of plant fossils. A comparison with present-day representatives of the lineages, when possible, represents one of the basic premises for paleontologic analysis. The anatomy found for *I. yansanae* was thus compared with that of present-day Gnetales, and the conditions

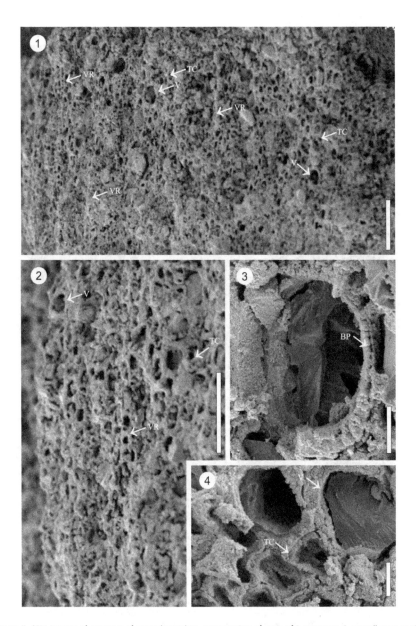

Figure 5. SEM images of anatomy of secondary xylem, cross-section of stem of *I. yansanae*: 1, overall cross-section showing vessels (V), tracheids (TC) and vascular raids (VR); 2, detail of View 1, showing vessels (V), tracheids (TC) and vascular raids (VR); 3, detail vessel (V) with bordered pits (BP); 4, detail of vessel (V) surround by tracheids (TC). Scale bars equal: 100µm in 1 and 2; 10µm in 3 and 4.

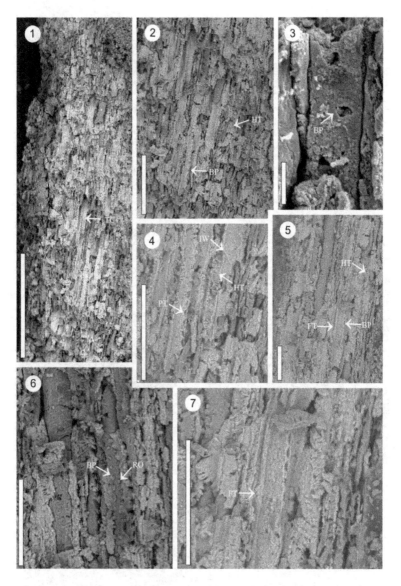

Figure 6. SEM images of anatomy of secondary xylem, tangential section of stem of *I. yansanae*: 1, overall tangential section showing tracheids (TC); 2, detail of View 1, showing tracheids with bordered pits (BP) and helical thickenings (HT); 3, detail of bordered pits (BP); 4, detail of View 2 showing pitted tracheids (PT), helical thickenings (HT) and inclined terminal walls (IW); 5, detail of View 1 showing tracheid with bordered pits (BP) and helical thickenings (HT) and fibre-tracheid (FT); 6, portion of tracheid with two rows of large alternate bordered pits (BP) with rounded openings (RO); 7, detail of View 5 showing long, thin fibre-tracheids (FT). Scale bars equal: 500μm in 1; 200μm in 2; 10μm in 3 and 100μm in 4, 5, 6 and 7.

of climate in which the fossils flourished were inferred. Thus, the gross morphology and anatomy of *I. yansanae* suggest that it grew in locations with a definite hydric deficit, at least during some seasons of the year; they also suggest, on the basis of other studies conducted in the Araripe Basin, such as the paleopalinological studies [5]. The ephemeral nature of the leaves may have been a defence of the plant to decrease the evaporative surface, with photosynthesis being the function of the new branches [24, 34, 35]. Another indication of the climatic conditions of hydric deficit during at least part of the year is the shortage of vessels elements in relation to the abundance of tracheids in the Gnetales. A study of the near vessellessness in *Ephedra*, [36] showed that variation in the xylem indicates an adaptation for improving the conduction of water. This is physiologically useful, but limited in availability, since tracheids are the vessels for the conduction of water because they avoid the formation of air bubbles or air embolisms. Moreover, present-day species of *Ephedra* with a low density of vessels per mm^2 have helical thickenings [37]. In the specimens studied here, the ratio of vessels per mm^2 was only 40 in comparison to the new-world and old-world species of *Ephedra* (1 to 291 and ≤20 to 338, respectively) [36, 37]; the helical thickenings of *I. yansanae* can also be interpreted as a consequence of the climate during the deposition of the Crato Member of the Santana Formation, which, as indicated above, would show that despite the system of lakes in the region, the climate was semi-arid.

The phylogenetic implications of the morphological and anatomical characteristics of *I. yansanae* exclude a relationship with any lineage other than that of Gnetales. Given its position in this order, it is closest to an ephedroid-like plant, although since the anatomical details of the seed envelope [27, 29, 31, 38, 39] and the pollen grains are unknown, we cannot prove that it is actually a member of the lineage of the *Ephedra*.

Character	Vegetative structures					Reproductive structures		
	Branches			Leaves	Roots			
Species	Type	Dimensions (cm)	Internode (cm)	Shape / Dimensions (mm)	Arrange / Venation	Female (mm)	Male (mm)	Seeds (mm)
Itajuba yansanae	Sympodial, dioecicous, longitudinally striated	54.5 x 0.75-1.15	22.5-9 x 4-0.5			3.6-5.3 x 2.6-2.8, two pairs of bracts		3 x 2.8, ovate surface ornamented with projections
Drewria potomacensis	Monopodial, longitudinally striated	Diameter 1-3	30	Oblong 10-20 x 2-6	Opposite Parallel with crossveins	Short loose spike arranged in groups of three		1-2.5 x 1-2 Flattened, narrowly ovate, apex acute, base rounded

Character	Vegetative structures					Reproductive structures		
	Branches			Leaves		Roots		
Species	Type	Dimensions (cm)	Internode (cm)	Shape / Dimensions (mm)	Arrange / Venation	Female (mm)	Male (mm)	Seeds (mm)
Ephedra archaeorhytido sperma	Opposite, erect, longitudinally striated		0.8-1.4 x 0.5-1	Caducous, triangular and acute, 5 x 2	Verticillate, parallel	Terminal, two to three pairs of bracts		Immature 1,5-4 x 1-1,6. Mature 7.5 – 2.2. Obovate – elongate-ovoid cuspidate apex, surface ornamented
Liaolaxia changii	Shrub(?) with opposite branch and longitudinally striated	8- 4 x 0.5-4				5-6 x 2.5-3 Sessile to pedunculate Compound, obovate, six pairs of bracts or more		0.9-1.1 x 0.3-0.7 Ovate
Liaoxia chenii	Longitudinally striated		9 x 0.1-0.3	Linear 20 x 1-2	Parallel	5-10 x 3-4 obovate, compound; two – six pairs of bracts		
Ephedra verticillata	Longitudinally striated	< 5				Single to five sessile structures		1.8 x 0.8, oval striated surface, with one pair of bracts
Siphonosperm um simplex				Linear	Opposite, parallel	Terminals obovate, without bracts, orthotropous ovule		

Character	Vegetative structures						Reproductive structures		
	Branches			Leaves		Roots			
Species	Type	Dimensions (cm)	Internode (cm)	Shape / Dimensions (mm)	Arrange / Venation		Female (mm)	Male (mm)	Seeds (mm)
Ephedra hongtaoi	Shrub	< 26	0. 7-4 x 0.4	Highly reduced		Tap root	3 x 2.3, spherical, two pairs of bracts		2.7 x 2.2 In pairs or single
?Gnetales	Shrub dioecicous, with opposite branch and longitudinally striated		2-13.5 x 1-2.5	Ovate-oblong, 3-7.5 x 1.5-5	Opposite	Closely together		Spikes (?)	
Cearania heterophylla	Herb – shrub, sympodial and longitudinally striated	"/>65.5		Linear lanceolate-ovate	Coriaceou s, parallel	Closely together	Single with many pairs of bracts		
Cariria orbiculiconifor mis	Herb – shrub, sympodial and longitudinally striated	16.5-0.5	4.4-4.5 – 0. 3-0.5	Oval to oval-orbicular and dorsiventrally flattened; decurrent base. Apex acute-obtuse or obtuse. 8-24 x 5-18mm	Parallel to the leaf margin		4-11 x1.2-1.5; orbicular compound strobili, a pair of sterile bracts	4-6 x ~1 individual pollen-producing structure with two sterile bract-like	
?Ephedra sp.	Shrub dioecicous, with opposite branch and longitudinally striate		2-13.5 x 1-2.5	Ovate-oblong, 3-7.5 x 1.5-5	Opposite	Closely together		Spikes (?)	

Table 1. Morphological characteristics of species associated with the Gnetales mentioned in the text (Modified from 4, 14, 25-30,).

Character	Stem	
	Cortical layer	Wood
Species		
Itajuba yansanae		Cross and tangential sections. Vessels and tracheids with alternate pittings, helical thickenings, fibre-tracheids and uniseriate vascular rays of procumbent cells
Cearania heterophylla	Uniform isodiametric and/or rectangular cells, helical thickenings, circular strands of supporting tissue	Transversal section. Vascular tissue, elongated cells with acute polar ends and helical thickenings, presence of pits arranged in single or two rows. Perforation plates and fibre-tracheids
Cariria orbiculiconiformis	Uniform parenchymatous and rectangular cells	Transversal section. Tracheids with helical thickenings and uniseriate pittings becoming biseriate at the polar ends. Fibre-tracheids

Table 2. Anatomical characteristics of species associated with the Gnetales mentioned in the text [Modified from 4, 12].

6. Conclusions

This paper has described a new species, *I. yansanae* on the basis of morphological and anatomical characteristics. It has been placed systematically in the order Gnetales, since it shares various characteristics with them, including the longitudinally striated stem, thickened nodes giving rise to varying numbers of branches; possibly ephemeral leaves, terminal female reproductive structures protected by two pairs of bracts and seeds externally ornamented with protuberances. The anatomy of the new species includes secondary wood consisting of vessels and tracheids with helical thickenings, fibre-tracheids and bordered pitting. This new species introduced one more taxon which contributes to a better understanding of the diversity of the Crato paleoflora during the Early Cretaceous period (Aptian-Albian).

Acknowledgements

The authors of this paper would like to acknowledge the collaboration of the Paleontological Museum of Monte Alto for lending the specimens, as well as by the important contributions of James A. Doyle, William DeMichele and the anonymous reviewer. They would also like to acknowledge the photographs of the specimens taken by Fabio C. Branco and the assistance of Linda Gentry El-Dash in the preparation of the English version of this text.

Author details

Fresia Ricardi-Branco[1], Margarita Torres[2], Sandra S. Tavares[1,4], Ismar de Souza Carvalho[3], Paulo G. E. Tavares[4] and Antonio C. Arruda Campos[4]

*Address all correspondence to: fresia@ige.unicamp.br

1 Departamento de Geologia e Recursos Naturais, Instituto de Geociências, Universidade Estadual de Campinas, Universidade Estadual de Campinas, Campinas, SP, Brazil

2 Centro Jardín Botánico, Facultad de Ciencias, Universidad de Los Andes. Mérida. Edo. Mérida La Hechicera, Venezuela

3 Departamento de Geologia, Instituto de Geociências, Universidade Federal do Rio de Janeiro, Cidade Universitária-Ilha do Fundão. RJ, Brazil

4 Museu de Paleontologia de Monte Alto, Prefeitura Municipal de Monte Alto, Praça do Centenário s/n - Centro de Artes - CEP: Monte Alto/SP, Brazil

References

[1] Assine M (2007) Bacia do Araripe. Boletim de Geociências da PETROBRAS, 15: 371-390.

[2] Coimbra J.C, Arai A, Careño, A.L (2002) Biostratigraphy of the Lower Cretaceous microfossils from the Araripe basin, northeastern Brazil. Geobios. 35: 687-698.

[3] Neumann V.H, Borrego A.G, Cabrera L, Dino R (2003) Organic matter composition and distribution through the Aptian-Albian lacustrine sequences of the Araripe Basin, northeastern Brazil. International Journal of Coal Geology. 54: 21-40.

[4] Kunzmann L, Mohr B.A.R, Bernardes-de-Oliveira M.E.C (2009) Cearania heterophylla gen. nov. et sp. nov., a fossil gymnosperm with affinities to the Gnetales from the Early Cretaceous of northern Gondwana. Review of Palaeobotany and Palynology. 158: 193–212. DOI:10.1016/j.revpalbo.2009.09.001

[5] Heimhofer U, Hochuli P.A (2010) Early Cretaceous angiosperm pollen from a low-latitude succession (Araripe basin, NE, Brazil). Review of Palaeobotany and palynology. 161: 105-126. DOI 10.1016/j.revpalbo.2010.03.010

[6] Fanton J.C.M (2007) Novas gimnospermas e possível angiosperma da Paleoflora Crato, Eocretáceo da bacia do Araripe, Nordeste do Brasil. Unpublished Master dissertation. Universidad Estadual de Campinas, 183p. Available: http://cutter.unicamp.br/document/results.php?words=fanton

[7]   Duarte L (1993) Restos de Araucariáceas da Formação Santana – Membro Crato (Aptiano), NE do Brasil. Annais da Academia Brasileira de Ciências. 65: 357-362.

[8]   Mohr B.A.R, Friis D.E.M (2000) Early angiosperms from the Lower Cretaceous Crato Formation (Brazil), a preliminary report. International Journal of Plant Sciences. 161 (6 Supplement): S155-S67.

[9]   Rydin C, Mohr B.AR, Friis E.M (2003) Cratonia cotyledon gen. et sp. nov.: a unique Cretaceous seedling related to Welwitschia. Proceeding of Royal Society of London B (Supplement) Biological Letters. 270: 1–4 .

[10] Kunzmann L, Mohr B.A.R, Bernardes-de-Oliveira M (2004) Gymnosperms from the cretaceous Crato Formation (Brazil). I. Araucariaceae and Lindleycladus (incertae sedis). Mitteilungen aus dem Museum für Naturkunde in Berlin, Geowissenschaftliche Reihe. 7: 155-174.

[11] Kunzmann L.B, Mohr A.R, Bernardes-de-Oliveira M, Wilde V (2006) Gymnosperms from the Early Cretaceous Crato Formation (Brazil). II Cheirolepidiaceae. Fossil Record. 9: 213-225.

[12] Kunzmann L, Mohr B.A.R, Wilde V, Bernardes-de-Oliveira M (2011) A putative gnetalean gymnosperm Cariria orbiculiconiformis gen. nov. et sp. nov. from the Early Cretaceous of Northern Gondwana. Review of Palaeobotany and Palynology. 165: 75-95. DOI: 10.1016/j.revpalbo.2011.02.005

[13] Dilcher D.A, Bernardes-de-Oliveira M, Pons, D, Lott T.A (2005) Welwitschiaceae from the Lower. Cretaceous of Northeastern Brazil. American Journal of Botany. 92: 1294–1310.

[14] Fanton J.C.M, Ricardi-Branco F, Dilcher D, Bernardes-de-Oliveira M. (2006a) New Gymnosperm related with Gnetales from the Crato Paleoflora (Lower Cretaceous, Santana Formation, Araripe basin, Northeastern, Brazil). Revista Geociências/ UNESP. 25: 205-210.

[15] Friis E.M, Crane P.R, Pedersen K.R (2011) The Early Flowers and Angiosperm Evolution. Cambridge University Press. Cambridge. pp 596.

[16] Fanton J.C.M, Ricardi-Branco F, Dilcher D, Bernardes -de-Oliveira M (2006b) Iara Iguassu, a new taxon of aquatic angiosperm from the Crato Paleoflora (Lower Cretaceous, Santana Formation, Araripe basin, Northeastern, Brazil). Revista Geociências/ UNESP. 25: 211-216.

[17] Mohr B.A.R, Bernardes-de-Oliveira M, Barale G, Ouaja M (2006) Palaeogeographic distribution and ecology of Klitzschophyllites, an Early Cretaceous angiosperm in Southern Laurasia and Northern Gondwana. Cretaceous Research. 27: 464–472. DOI: 10.1016/j.cretres.2005.08.001

[18] Mohr B.A.R, Bernardes-de-Oliveira M, Loveridge R.F (2007) The macrophyte flora of the Crato Formation. In: Martill D.M, Bechly G, Loveridge R.F, editors. The Crato

fossil Beds of Brazil: window into an ancient world. Cambridge University Press, Cambridge. pp. 537–565.

[19] Mohr B.A.R, Bernardes-de-Oliveira M, TAYLOR D.W (2008) Pluricarpellatia, a nymphaealean angiosperm from the Lower Cretaceous of Northern Gondwana (Crato Formation, Brazil). Taxon, 57: 1147–1158.

[20] Martill D.M (2007) The geology of the Crato Formation. In: Martill D.M, Bechly G, Loveridge R.F, editors. The Crato fossil beds of Brazil: window into an ancient world. Cambridge University Press. pp. 2-24.

[21] Frey W (2009) Subkingdom Embryobionta Cronquist, Takht. & W. Zimm. In: Frey W, editor. Syllabus of plant families: A. Engler's Syllabus der Pflanzenfamilien 3. Bryophytes and seedless vascular Plants. Gebr. Borntraeger Verlagsbuchhandlung. Germany. pp. 6-8

[22] Kubitzki K (1990) Gnetaceae: with the single Order Gnetales. In: Kramer K.U, Green P.S, editors. Pteridophytes and Gymnosperms. Springer-Verlag, pp. 378-391.

[23] Crane P.R, (1996) The fossil history of the Gnetales. International Journal of Plant Sciences. 157 (6 Supplement): S50-S57.

[24] Carlquist S (1996) Wood, bark and stem anatomy of Gnetales: a summary. International Journal of Plant Sciences. 157 (6 Supplement): S58-S76.

[25] Crane P.R, Upchurch G.R (1987) Drewria potomacensis gen. et sp. nov., an Early Cretaceous member of Gnetales from the Potomac Group of Virginia. American Journal of Botany. 11: 1722-1736.

[26] Yang Y, Geng B.Y, Dilcher Dchen, Z.D, Lott T. A (2005) Morphology and affinities of an Early Cretaceous Ephedra (Ephedraceae) from China. American Journal of Botany. 92: 231-341.

[27] Rydin C, Wu S, Friis E.M (2006a) Liaoxia (Gnetales): ephedroids from the Early Cretaceous Yixian Formation in Liaoning, northeastern China. Plant Systematics and Evolution. 262: 239–265.

[28] Wang X, Zeng S, (2010) Whole fossil plants of Ephedra and their implications on the morphology, ecology and evolution of Ephedraceae (Gnetales). Chinese Science Bulletin, 55: 1511-1519.

[29] Rydin C, Friis E.M (2010) A new Early Cretaceous relative of Gnetales: Siphonospermum simplex gen. et sp. nov. from the Yixian Formation of Northeast China. BMC Evolutionary Biology. 10:183. Available: http://www.biomedcentral.com/1471-2148/10/183.

[30] Cladera G, Del Fueyo G.M, Villar de Soane L, Archangelsky S (2007) Early Cretaceous riparian vegetation in Patagonia, Argentina. Revista del Museo Argentino de Ciéncias Naturales. 9: 49-58.

[31] Rydin C, Pedersen K.R, Crane P.R, Friis E.M (2006b) Former diversity of Ephedra (Gnetales): evidence from Early Cretaceous seeds from Portugal and North America. Annals of Botany. 98: 123–140.

[32] Ickert-Bond S.M, Rydin C (2011) Micromorfology of the seed envelope of Ephedra L. (Gnetales) and its relevance for the timing of evolutionary events. International Journal of Plant Sciences. 172: 36-48. DOI: 10.1086/657299

[33] Friis E.M, Pedersen K.R, Crane P.R (2009) Early Cretaceous mesofossils from Portugal and eastern north America related to the Bennettitales-Erdtmanithecales-Gnetales Group. American Journal of Botany. 96: 252–283. DOI: 10.3732/ajb.0800113

[34] Carlquist S (1989) Wood and bark anatomy of the new world species of Ephedra. Aliso. 12: 441-483.

[35] Price R (1996) Systematics of the Gnetales: a review of morphological and molecular evidence. International Journal of Plant Sciences. 157 (6 Supplement): S40-S49.

[36] Carlquist S (1988) Near-vessellessness in Ephedra and its significance. American Journal of Botany. 75: 598-601.

[37] Carlquist S (1992) Wood, bark and pith anatomy of the old world species of Ephedra and summary for the genus. Aliso, 13, 255-295.

[38] Rydin C, Khodabandeh A, Endress P. K (2010) The female reproductive unit of Ephedra (Gnetales): comparative morphology and evolutionary perspectives. Biological Journal of the Linnean Society. 163: 387-430.

[39] Yong Y, (2010) A review on Gnetalean megafossils: Problems and perspectives. Taiwania. 55: 346-354.

Permissions

The contributors of this book come from diverse backgrounds, making this book a truly international effort. This book will bring forth new frontiers with its revolutionizing research information and detailed analysis of the nascent developments around the world.

We would like to thank Yuanzhi Zhang and Pallav Ray, for lending their expertise to make the book truly unique. They have played a crucial role in the development of this book. Without their invaluable contribution this book wouldn't have been possible. They have made vital efforts to compile up to date information on the varied aspects of this subject to make this book a valuable addition to the collection of many professionals and students.

This book was conceptualized with the vision of imparting up-to-date information and advanced data in this field. To ensure the same, a matchless editorial board was set up. Every individual on the board went through rigorous rounds of assessment to prove their worth. After which they invested a large part of their time researching and compiling the most relevant data for our readers. Conferences and sessions were held from time to time between the editorial board and the contributing authors to present the data in the most comprehensible form. The editorial team has worked tirelessly to provide valuable and valid information to help people across the globe.

Every chapter published in this book has been scrutinized by our experts. Their significance has been extensively debated. The topics covered herein carry significant findings which will fuel the growth of the discipline. They may even be implemented as practical applications or may be referred to as a beginning point for another development. Chapters in this book were first published by InTech; hereby published with permission under the Creative Commons Attribution License or equivalent.

The editorial board has been involved in producing this book since its inception. They have spent rigorous hours researching and exploring the diverse topics which have resulted in the successful publishing of this book. They have passed on their knowledge of decades through this book. To expedite this challenging task, the publisher supported the team at every step. A small team of assistant editors was also appointed to further simplify the editing procedure and attain best results for the readers.

Our editorial team has been hand-picked from every corner of the world. Their multi-ethnicity adds dynamic inputs to the discussions which result in innovative outcomes. These outcomes are then further discussed with the researchers and contributors who give their valuable feedback and opinion regarding the same. The feedback is then collaborated with the researches and they are edited in a comprehensive manner to aid the understanding of the subject.

Apart from the editorial board, the designing team has also invested a significant amount of their time in understanding the subject and creating the most relevant covers. They scrutinized every image to scout for the most suitable representation of the subject and create an appropriate cover for the book.

The publishing team has been involved in this book since its early stages. They were actively engaged in every process, be it collecting the data, connecting with the contributors or procuring relevant information. The team has been an ardent support to the editorial, designing and production team. Their endless efforts to recruit the best for this project, has resulted in the accomplishment of this book. They are a veteran in the field of academics and their pool of knowledge is as vast as their experience in printing. Their expertise and guidance has proved useful at every step. Their uncompromising quality standards have made this book an exceptional effort. Their encouragement from time to time has been an inspiration for everyone.

The publisher and the editorial board hope that this book will prove to be a valuable piece of knowledge for researchers, students, practitioners and scholars across the globe.

List of Contributors

Alfons Callado, Pau Escribà, José Antonio García-Moya, Jesús Montero, Carlos Santos, Daniel Santos-Muñoz and Juan Simarro
Agencia Estatal de Meteorología (AEMET), Spain

Nicolas Hautière, Raouf Babari and Eric Dumont
Université Paris-Est, Institut Français des Sciences et Technologies des Transports, de l'Aménagement et des Réseaux, France

Jacques Parent Du Chatelet
Météo France, France

Nicolas Paparoditis
Université Paris-Est, Institut Géographique National, France

Martina Tudor, Stjepan Ivatek-Šahdan, Antiono Stanešić, Kristian Horvath and Alica Bajić
Croatian Meteorological and Hydrological Service, Croatia

Sarah N Collins, Robert S James, Pallav Ray, Katherine Chen, Angie Lassman and James Brownlee
Department of Marine and Environmental Systems, Florida Institute of Technology, Melbourne, FL, USA

Zhong Zhong, Yijia Hu, Xiaodan Wang and Wei Lu
College of Meteorology and Oceanography, PLA University of Science and Technology, Nanjing, China

Diandong Ren and Mervyn J. Lynch
Department of Imaging and Applied Physics, Curtin University, Australia

Lance M. Leslie
School of Meteorology, University of Oklahoma, USA

Nagayoshi Katsuta and Shin-ichi Kawakami
Faculty of Education, Gifu University, Japan

Ichiko Shimizu
Department of Earth and Planetary Science, Graduate School of Science, University of Tokyo, Japan

Mineo Kumazawa and Masao Takano
Graduate School of Environmental Studies, Nagoya University, Japan

Herwart Helmstaedt
Department of Geological Science, Queen's University, Canada

Sueli Yoshinaga Pereira, Melina Mara de Souza, Fresia Ricardi-Branco and Renato Zázera Francioso
Institute of Geosciences, State University of Campinas, Brazil

Paulo Ricardo Brum Pereira
Forestry Institute, Sao Paulo State Environmental Secretariat, Brazil

Fabio Cardinale Branco
EnvironMentality – Conceitos Ambientais LTDA, Brazil

Pothiraj Prabu and Baskaran Rajagopalan
Department of Industries and Earth Sciences, Tamil University, Thanjavur, Tamilnadu, India

Fresia Ricardi-Branco
Departamento de Geologia e Recursos Naturais, Instituto de Geociências, Universidade Estadual de Campinas, Universidade Estadual de Campinas, Campinas, SP, Brazil

Margarita Torres
Centro Jardín Botánico, Facultad de Ciencias, Universidad de Los Andes, Mérida, Edo, Mérida La Hechicera, Venezuela

Ismar de Souza Carvalho
Departamento de Geologia, Instituto de Geociências, Universidade Federal do Rio de Janeiro, Cidade Universitária-Ilha do Fundão. RJ, Brazil

Paulo G. E. Tavares and Antonio C. Arruda Campos
Museu de Paleontologia de Monte Alto, Prefeitura Municipal de Monte Alto, Praça do Centenário s/n - Centro de Artes - CEP: Monte Alto/SP, Brazil

Sandra S. Tavares
Departamento de Geologia e Recursos Naturais, Instituto de Geociências, Universidade Estadual de Campinas, Universidade Estadual de Campinas, Campinas, SP, Brazil
Museu de Paleontologia de Monte Alto, Prefeitura Municipal de Monte Alto, Praça do Centenário s/n - Centro de Artes - CEP: Monte Alto/SP, Brazil